Patios & Decks
How to Plan, Build & Enjoy

by Michael Landis & Ray Moholt

HPBooks

HPBooks

Publishers
Bill and Helen Fisher

Executive Editor
Rick Bailey

Editorial Director
Randy Summerlin

Editor
Jim Barrett

Art Director
Don Burton

Book Design
Kathleen Koopman

For Horticultural Publishing Co. Inc.

Executive Producer
Richard M. Ray

Coordinating Editor
Lance Walheim

Associate Editor
Michael MacCaskey

Production Editor
Kathleen S. Parker

Contributing Editor
Susan Chamberlin

Illustrations
Roy Jones

Major Photography
Michael Landis

Additional Photography
Western Wood Products Association

Photographic Assistant
Richard B. Ray

Published by HPBooks
P.O. Box 5367
Tucson, AZ 85703
602/888-2150
ISBN: 0-89586-162-3
Library of Congress Catalog Card
Number: 82-84041
© 1983 Fisher Publishing Inc.
Printed in U.S.A.

Acknowledgments

We would like to thank the following people and firms for their valuable assistance:

Gary Blum, Ireland's Landscaping Inc., East Norwich, NY
Roy Foster, Van Dusen Botanical Display Garden, Vancouver, British Columbia, CAN
Jim Gibbs, Green Bros. Landscape Co. Inc., Smyrna, GA
Goldberg & Rodler Inc., Huntington, NY
Rudi Harbauer, Atlantic Nursery & Garden Shop Inc., Freeport, NY
Historic New Orleans Collection, New Orleans, LA
Landscape Architecture Magazine, Louisville, KY
Wayne Leong, St. Helena, CA
Longue Vue House and Gardens, New Orleans, LA
James L. Loper, Louisville, KY
David Madison, New York, NY
Gary Martin, Landscape Innovations, Inc., Melville, NY
Jim Olson, Olson & Walker, Architects, Seattle, WA
Tom & Jerry Enterprises, Inc., New York, NY
J. Hart and H. Tagami, Kaneohe, HA
Warner, Walker & Maly, P.C., Landscape Architects and Planners, Portland, OR
Western Wood Products Association, Portland, OR
Wimmer, Yamada & Associates, San Diego, CA

We would also like to thank:

Basalt Products, Napa, CA
Jack Cohen, Vancouver, British Columbia, CAN
Dr. William and Christine Cress, Troy, MI
Walter Doty, Los Altos, CA
Ted Efstratis, Sacramento, CA
Mrs. Mark Ellis, West Vancouver, British Columbia, CAN
The Flood Company, Hudson, OH
Peggy Fortier, New Orleans, LA
Mr. and Mrs. Paul Godfrey, New Orleans, LA
Mr. and Mrs. Dana Jones, College Park, GA
J.K. Hutchinson, Palo Alto, CA
Harry Kilpatrick, Browns Valley, CA
Mary Landis, Rutherford, CA
Mr. and Mrs. Terry Larkin, St. Helena, CA
Ghitta and Ulric Lejeune, Lions Bay, British Columbia, CAN
Mr. and Mrs. Hugh Martin, Vancouver, British Columbia, CAN
Don McCarter, New Orleans, LA
John Meath, Benicia, CA
Theodore Metalios, Jackson Heights, Queens, NY
Jean Micheals, St. Helena, CA
Mortex Mfg. Co. Inc., Tucson, AZ
Lowel Nesbith, New York, NY
Mr. and Mrs. Johnathan Parker, Vancouver, British Columbia, CAN
Mr. and Mrs. Victor Pavel, Honolulu, HA
Fred Rea, Palo Alto, CA
John Reinhart, Laguna Hills, CA
Mrs. A.W. Robertson, Vancouver, British Columbia, CAN
Fritzy Schultz, St. Helena, CA
Hans Sumpf, Fresno, CA
Tri City Concrete, Fairfield, CA
Eillen Traverse, Piedmont, CA
Jim Weirich, New Orleans, LA
Mrs. Edmond Wingfield, New Orleans, LA

The landscape design of James L. Loper, pages 34-36, originally appeared in the March 1981 issue of *Landscape Architecture* magazine.

Contents

Cover Photograph:
The pleasing geometric lines of this multilevel deck and patio point the way to an outdoor dining area. A natural-gas firepit in the table center provides warmth for diners on chilly evenings. Design: Gerald B. Fischer of Stone-Fischer Associates, Landscape Architects, Del Mar, CA. Photo by Richard Fish.

Outdoor Living

Ask homeowners why they want a patio or deck and you will almost *always* get the same answer. They want to spend more time outdoors in comfort. Ask them what they want to do once they're outdoors and you may *never* get the same answer!

Outdoor living has as many variations as the people who enjoy it. Cooking, eating, playing, entertaining, relaxing, gardening—the list of outdoor activities goes on and on. It would be a mistake to think of a patio or deck as a surface on which only these activities occur. A properly designed patio or deck can be as useful as any room in your house.

INCREASE YOUR
LIVING SPACE

A patio or deck is an economical way of extending living space. It can be any size or shape and positioned almost anywhere. Roof decks are becoming a popular solution to the problem of limited space in urban areas. Patios and decks can turn awkward areas into comfortable outdoor living space.

A patio or deck can combine the fresh feeling of being outdoors with the security and convenience of being indoors. A small alcove next to a bedroom provides a cool place to sleep on a hot summer night. Kitchens are usually a center of family activity

in summer. An adjacent patio or deck for outdoor eating opens up valuable room indoors.

A patio or deck can be designed for all-weather enjoyment, or it can be completely exposed to the elements. Patios and decks are usually much less expensive than room additions. Indoor activities are rarely disturbed during patio and deck construction, which is so often the case with most major home additions.

GIVE AN OLD HOUSE
NEW LIFE

Old homes can be given a modern look with the addition of patios or decks. From new entryways to multi-level extensions, patios and decks add new utility and enjoyment to older property.

When stripped of shrubbery, the tall foundations of older homes often break the smooth visual transition from home to garden. Raised decks

Left: Traditional red brick patio and walk look natural in this lush garden setting. For details on installing brick patios, see page 104.

Tile patio brightens up older home, lends a formal appearance to the yard. Wooden steppingstones in foreground help define pathway across the lawn.

can ease this transition and create more-usable planting areas.

Energy efficiency is now an everyday consideration. Many older homes were not designed with energy savings in mind. Strategically placed patio and deck overheads and arbors can cut home cooling and heating costs.

SOLVE LANDSCAPING PROBLEMS

Patios and decks can create usable space out of "unusable" space. Decks can be raised over areas with poor soil or water drainage. They can also be built over steep terrain without disturbing the natural contours of the slope.

Textured patios give cluttered or unused areas new grace and order. The wide array of patio surfacing materials allows you to add visual interest to even the smallest, least-promising spots.

Decks allow you to make the most of natural features that already exist in your landscape. They can be built around large trees so you can enjoy cooling shade and still provide necessary water, nutrients and air to roots. Deck edges can be contoured to follow the shape of large rocks or boulders, as shown on page 153.

ENHANCE THE APPEARANCE OF INDOOR ROOMS

A patio or deck creates an inviting transitional space between indoors and outdoors. When viewed from indoors, a patio links the yard or garden with the house. It makes the house interior seem larger and more open.

An outdoor patio or deck visible from indoors is an invitation outside to enjoy a spectacular view or proud garden. At the same time, it brings the beauty of the plants around the patio or deck indoors, making the room feel less confined.

PROVIDE OUTDOOR PRIVACY

You can build a patio or deck to satisfy your own tastes in outdoor enjoyment. For many people, this means privacy and solitude—a chance to step outside the commotion of everyday life.

Patios and decks can be used to sidestep everyday activity as effectively as they can be used to supplement it. As garden floors, they signify a place of intended purpose. Build a patio or deck in an out-of-the-way

A small deck in an unused corner of the yard makes a secluded retreat.

Patio and wall match the house siding to create architectural unity.

Tile patio links indoor and outdoor living areas. Many kinds of tile can be used both indoors and out.

Decks easily provide usable space on steep hillside lots. This one is connected to the house.

grove of trees to create a quiet retreat, or place it in a secluded side yard away from activities.

Complement patios and decks with hedges, fences and screens to block out unwanted views or viewers. Even in a busy neighborhood, you can enjoy outdoor privacy.

How To Use This Book

There are many different deck and patio designs that will fit your house and outdoor lifestyle. There are numerous shapes, sizes, textures, materials and accessories you can choose. Once you choose and design a patio or deck that's right for you, there is a *right* way to build it.

This book offers a blend of design concepts, idea-inspiring photography and sound building instructions.

The chapter, "Planning Your Patio Or Deck," beginning on page 9, gives you information that will help you design your patio or deck. Your participation in the design process is important. You can get the patio or deck you want regardless of whether you do the building or not. The chapter contains hints on how to assess the esthetic and physical qualities of your property. You can learn how to solve difficult problems such as sun and shade control, poor drainage or steep slopes. You'll learn to recognize *microclimates*, or hot and cool areas, around your home so you can position a patio or deck in the most-comfortable location.

If your possible construction sites are limited, this chapter shows you how to use physical features such as hedges, fences, screens and arbors to make a difficult site more comfortable for users. There are tips on how to incorporate trees, rocks and other natural features into your design and how to put your ideas on paper.

The last section of the design chapter includes interviews with professional designers and architects. These experts reveal how they approach patio and deck design with the needs of their client in mind. The section includes photographs of the projects the professionals discuss.

The chapter, "Patio And Deck Ideas," beginning on page 45, is a photographic survey of deck and patio ideas. It's organized into categories such as *indoor-outdoor patios and decks, roof decks, courtyards, overheads, railings* and *steps.*

Seeing what others have done can help you visualize your own patio or deck. The photographs show you how materials, structures, colors and patterns blend together and how they affect the house. You will see that problems can be handled in many ways. Most important, you will see a lot of creative design, construction and use of materials that will take you one step closer to your own personalized outdoor-living area.

The last two chapters, "Patio Builder's Guide" and "Deck Builder's Guide," beginning on pages 101 and 125, give precise instructions on patio and deck construction. They contain detailed, step-by-step illustrations that lead you from buying materials and tools through proper installation.

These chapters will help you build a structurally sound, visually pleasing patio or deck.

There are many ways to build a deck. The right tools, materials and help can make the job much easier and problems less frequent. The chapter on deck construction is designed to let you see most of the alternatives before you drive the first nail. Problems are solved on paper and construction goes smoothly.

Patio construction is more straightforward. It's mostly a question of materials. Each masonry material requires specialized building techniques. But more important, individual materials look, wear and feel different. This chapter not only gives basic techniques for installing these materials, but also helps you compare durability, ease of installation and appearance.

This spacious patio is in scale with its surroundings. A large patio can provide room for a number of outdoor activities.

This rooftop patio and deck allow the owner to sit among the treetops. Consider a rooftop patio if yard space is limited.

Planning Your Patio Or Deck  2

Think about the site you have and how you want to use your patio or deck. Do you entertain frequently? Do you have young children with lots of friends? Do you enjoy outdoor work, or hope to avoid it? Will your patio or deck be used during the day or evening?

These are the kinds of questions an expert designer asks. They are questions you must ask yourself. The answers will help you begin your design.

Many building books recommend planning on paper. This is one approach. If you have trouble visualizing three-dimensional spaces on paper, start outside.

How And Where To Begin

Begin with a study of your yard. If you already have an accurate *base map* of your property, take it outside with you. Note your observations directly on it. If you prefer working with an existing base map as your guide, see pages 11-12 for instructions on making one.

Another good approach is to spend time in your yard with a notebook and pencil, recording your observations and making sketches. Go at different times of the day and in different sea-

sons, if possible. Note wind direction, sunny and shady areas and shape of the land. Determine which views to block—neighbors' windows, roads, storage areas—and which views to accent. Determine the rainwater-drainage pattern. Carefully note low spots that collect water. For more information on drainage problems, see pages 24-26.

SUN

The intensity of sunlight throughout the day and seasons of the year is an important design factor. Sunlight helps determine how you organize your outdoor-living area.

In cool climates where warm, sunny days are few, seek ways to make the most of sunlight. Concrete, tile, brick and asphalt surfaces absorb heat during the day and release it at night.

Shade is important where summers are long and temperatures are high. Block sun with trees, arbors or overhead screens. Keep heat-absorbing paving to a minimum. Remember that dark-colored paving absorbs more heat than light-colored paving.

In many climates, adjustable overheads are useful in controlling the amount of sunlight that falls on the patio or deck. Some overheads have

Left: A hedge of dark greenery provides the perfect backdrop for this elegant small deck and display platform.

U-shaped bench and vine-covered trellis separate conversation and entertaining areas. When planning your patio or deck, consider how you will define activity areas.

adjustable louvers for this purpose. Others are simply frames that support removable bamboo or canvas screens.

Deciduous trees provide shade during the summer and admit sunlight during the winter. For more information on controlling sun and shade, see pages 14-15.

WIND

A cool breeze on a hot afternoon is wonderful. A chilly wind that moves everyone inside is not.

Most houses are big, effective windbreaks. Take advantage of sheltered spots around yours. Consider using plants for additional windbreaks or installing vertical windscreens of glass or wood. If breezes spill over the roof onto a potential patio area, an overhead screen or roof may be necessary.

When you plan the deck or patio, provide one area that takes advantage of cool breezes and another spot that is sheltered from harsh winds. One way to achieve this dual characteristic is to wrap the patio or deck around a corner of the house.

Learn about wind around your house by walking around the yard at different times of day. Ask neighbors, keeping in mind that winds around their houses may be different than around yours. Check the direction trees and shrubs lean. Call local meteorologists and ask about the direction of prevailing winter and summer winds. For more information on dealing with wind problems, see pages 16-17.

SURROUNDING FEATURES

Take nature's cue. Existing trees, rocks or interesting views are the designer's greatest assets. If you have any of these or other natural features, capitalize on them. Increasing your outdoor living space in the least-obtrusive way is good design.

Decks are easily built around trees and rocks. They are an excellent way to preserve existing trees on the property. There are two types of masonry surfaces that you can use around trees. One is loose gravel. The other is bricks set on a bed of sand. Both surfaces allow air and water to reach tree roots.

Gentle slopes may be accentuated by a multilevel deck or terrace. Patios of local stone or aggregate harmonize with the environment. Blend your design with nature to produce a pleasing effect.

An indoor-outdoor connection is always pleasing. It is even more attractive when natural features like these rocks are incorporated into the design.

Study not only your own property but other lots in the neighborhood. Notice building materials, construction details, native plants and the character of the land itself. You will get ideas for your own deck or patio.

INDOOR-OUTDOOR CONNECTION

Design your patio or deck as an outdoor extension of indoor living space. Think about this as you study your property. French doors or sliding-glass doors leading to a deck or patio dramatically increase the sense of spaciousness inside a house. Glass doors make rooms lighter and more airy, and they add adjacent floor space for entertaining or relaxing.

Use the same flooring material inside and out to link the two areas. Tile is especially well-suited for this use. For information on tile patios, see page 116.

Indoor-outdoor connections need not be large or elaborate. A small, plain patio or deck off the kitchen or bedroom can be enjoyable. Maybe you would like to extend your children's playroom with a small, secure patio, or add a deck to enlarge a second-story studio. There are many possibilities, so keep the indoor-outdoor connection in mind as you study your yard.

Also consider space requirements as you look at potential sites. Your

Existing trees offer many opportunities. Here, trees create a shady retreat. The patio could also have been placed next to the house as a platform for viewing the trees.

patio or deck must accommodate you, your children and friends. Plan space for all the activities that will take place on the patio or deck. For more information on planning spaces, see pages 18-20.

LOCAL CODES

Before you get too far into planning, get to know local building codes. They not only dictate building practices, but also govern patio or deck size, stair and railing construction and placement within the property.

Most communities have required setback distances from property lines and streets. Most codes require safety railings on decks above ground level. There are codes governing height of fences and screens around the patio or deck. Check with your local building department for requirements.

If you live outside an incorporated area, building codes may not be as strict. Avoid wasted time during planning by checking with your local building department first. Sometimes building officials will even help you with plans.

Your plans will have to be approved by the building department before you can get a building permit. If your plans don't meet local codes, you'll have to rework them so they do. One exception would be if you get a *zoning variance* from city or county officials. Permits and variances are discussed on page 30.

Making Plans

Your notebook is full of observations and ideas. You have checked your local building codes. Now it's time to start making decisions.

MAKING A BASE MAP

A base map is a scale drawing of your site as if viewed from directly overhead. It is an exact plan that shows all important existing features of your property.

This is probably the most useful plan you can make. It is a *site inventory* available for future reference and constant refinement. You can use it to estimate building materials. Equally important, the base map becomes the record of your ideas and your property's development. The map should include:

- Overall lot dimensions and lot orientation to north.
- Location of the house on the lot. This should be a floor plan showing the location of rooms, doors and windows.
- Existing trees, important plants and outdoor structures.
- Locations of hose bibs, exterior electrical outlets and utility meters. Include locations of underground utility lines if these will be affected by construction.

A sample base map is shown on page 12.

Creating a base map from scratch is a lot of work. You will save hours of measuring if you can locate a *deed map, assessor's parcel map, contour map, architect's drawings* or *house plans.*

Deed maps or assessor's parcel maps show lot dimensions and orientation of the property to the surrounding area. If you don't have one, check with the city clerk, county recorder or the bank or title company carrying the mortgage on your house. Deed maps often depict a large number of lots or parcels. Don't be surprised if you need a magnifying glass to read your lot dimensions on the map.

A contour map is important for hillside sites. The map shows natural drainage areas on your property. It may be required to get a building permit. Made by a survey company or landscape architect, the map shows contour lines that indicate *grade*, or ground-level changes on your site. Contour lines typically indicate grade

When you are striving for an indoor-outdoor connection, architecture is important. Examine your house for features that may affect your design. This house's architecture offers no design clues. The striking patio-arbor addition has actually improved the facade.

Consider existing architecture when designing your deck or patio. It was necessary to build this deck up to the house floor level and add doors to the adjacent room. Staining the deck the same color as the house makes it look like part of the original construction. A staircase connects the deck with a lower patio.

DRAWING A BASE MAP

STEP 1: Obtain a deed map or assessor's parcel map. The map should show the dimensions of your property. Architect's drawings or house plans will be of help in drawing your base map.

STEP 2: Reproduce property line information exactly on a 24x36" piece of 100% rag tracing paper with graph lines on it. This paper is sold in some stationary stores and stores that stock engineer's and architect's supplies. Use a scale of 1/4 inch equals 1 foot, or 1/8 inch equals 1 foot. This will be your base map.

STEP 3: Place tracing paper over the base map. Go outside with a helper and a long tape measure. Make a *site analysis.* Locate all important features in the yard by measuring their distances from known points. Transfer these measurements to the overlay. Make rough notes only. Do not try to make a detailed drawing at this point. Remove this overlay when you have completed your analysis.

STEP 4: Place a fresh overlay on the base map. Refer to your analysis. Accurately locate all features using proper scale. When the overlay is complete, place it *beneath* the base map. Trace overlay information onto base map to complete it.

The design process begins here. It culminates on page 28 with a sample design solution for this fictional property.

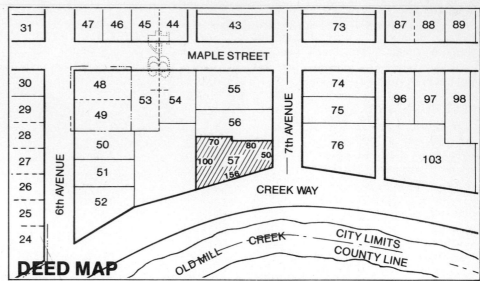

DEED MAP

A deed map or parcel map shows the basic dimensions of your property.

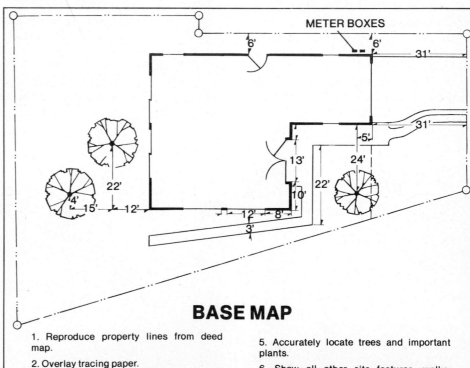

BASE MAP

1. Reproduce property lines from deed map.

2. Overlay tracing paper.

3. Locate house squarely by measuring from known property lines—streets, fences, surveyor's stakes. Make rough notes.

4. Show all doors and windows so indoor-outdoor relationships will be clear.

5. Accurately locate trees and important plants.

6. Show all other site features—walks, driveways, walls, fences, meters, fountains, rock outcroppings and so forth.

7. Remove the overlay. Accurately draw information to scale on a fresh overlay. Place the new overlay beneath the base map and transfer the information to the map.

Completed base map shows all important features on the property.

changes in increments of 1, 2, 5 or 10 feet.

Architect's drawings of your house show the site plan, floor plan and other important details. You should verify the plan's indicated location of the house on the property. Take actual measurements from the corners of the house to the property lines. Builders sometimes shift the house location by several feet from what is indicated on the plans.

Make your base map on a large sheet of tracing paper with graph lines on it. The best size is a 24x36" sheet with a 1/8-inch or 1/4-inch grid.

When you make your base map, each graph square will represent one square foot. If you plan to make copies of the map, the graph lines should be "fade-out" blue. These lines won't reproduce when the map is copied. If you plan to make blueprints from the map, use 100% rag

paper, available at stores that carry engineer's or architect's supplies. At the same time, buy some tracing paper of the same size. To draw the base map, follow the steps outlined above.

SITE ANALYSIS

Once the base map is completed, lay a piece of tracing paper over it. Transfer notes from your notebook to

the tracing paper overlay. Make your *site analysis* on this tracing paper.

The site analysis includes your observations on sunlight and wind patterns, views, slopes, drainage problems and so on. Include any code-required setbacks that may affect construction.

Lay another piece of tracing paper over this sheet. Begin experimenting with designs for your patio or deck.

BUBBLE PLAN

A *bubble plan* is nothing more than a casual sketch or doodle. It evolves naturally from a combination of your base map and notes. It is the best way to experiment with ideas without getting bogged down in specifics.

Lay tracing paper over the base map. Roughly sketch in possible patio and deck areas. You can quickly explore dozens of ideas this way.

Bubble plans are no beginner's crutch. Professional designers depend on this give-and-take process. Each plan attempted is a step toward the final design. Each one discarded is a record of your decisions. A sample bubble plan is shown below right.

SITE PLAN AND WORKING DRAWINGS

These plans represent the final stages of design. The *site plan,* sometimes called a *plot plan,* is carefully drawn and detailed. It is a combination of the base map and bubble plan. All new features are shown in their final form and in proper scale.

Working drawings are important if you decide to hire a contractor. These include the site plan and any detail drawings, such as elevation drawings for fences, patio overheads or other structures.

A set of plans saves money because contractors can see exactly what you have in mind. A contractor uses plans to give you accurate cost estimates. Time will not be wasted because of poor communication.

When added to a contract, working drawings become legal documents. Contractors agree to build from your plans *exactly as shown.* They cannot substitute cheaper materials or omit any features. If you have working drawings, your written agreement with the contractor can be much simpler. For more information on dealing with contractors, see page 30.

Working drawings are also useful if your patio or deck plan includes custom-design features. The drawings are more often required for decks than for patios. They are necessary for getting building permits for raised decks or retaining walls more than 3 feet high.

You can draw your own site plan and working drawings if you include the necessary information. A sample site plan is shown on page 28. Working drawings for decks are discussed on page 125. If your patio or deck con-

struction is simple and you intend to do the work yourself, detailed plans aren't absolutely necessary. You can arrange with a landscape architect, architect or engineer to supply final working drawings. They will work

from your base map and bubble plan. This method saves money in designer's fees. You can then give the plans to a contractor or you can arrange for your designer to deal with contractors directly.

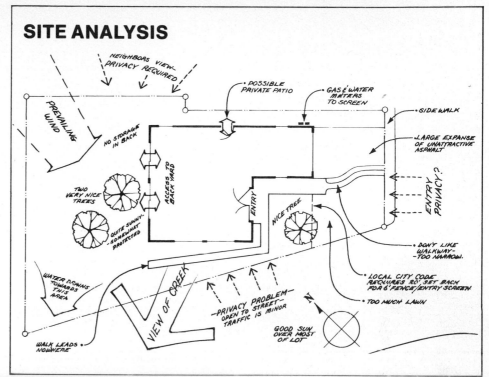

SITE ANALYSIS

Every site is different. Careful examination of your property reveals factors that will affect your design. Set aside preconceived notions and observe. Record your observations on the site analysis.

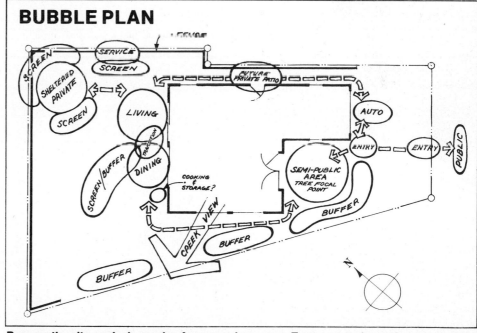

BUBBLE PLAN

Remove the site analysis overlay from your base map. Tape a new piece of paper over the base map. Think about activities that will take place on the patio or deck. Consult your site analysis notes. Consider how activities relate to the actual property. Sketch bubble plans to represent basic areas for activities and uses. Let ideas flow quickly. Overlay as many pieces of tracing paper as necessary. The bubble plan illustrated here shows ideas leading to the final design of this fictional property, shown on page 28. A real bubble plan need not be drawn as neatly as this one.

Take Advantage Of Microclimates

Microclimates occur in small areas affected by large physical objects such as trees or houses. These objects cause the climate immediately around them to be slightly different from the general climate of an area. A cool, shady spot during a hot, summer day or glaring heat from the side of a house are examples of microclimates.

Understanding the effects of sun and wind around your home is important. This knowledge will enable you to take advantage of microclimates and determine the best site for your outdoor-living area.

There are ways to change the existing microclimate of a site. By controlling the way sun or wind affects the site, you can make your patio or deck more comfortable.

Sunlight is usually the most important factor. The sun's position in the sky changes daily and seasonally. To determine shade patterns, make observations at different times of the day and different times of the year. The shade patterns discussed here apply to homes located in the Northern Hemisphere.

The north side of buildings in the Northern Hemisphere is shaded. Use this principle to your advantage in hot climates. In cooler climates, north-facing patios can be extended to a sunny spot for sunbathing. Move furniture to take advantage of house-shadow locations at various times of the year. This light-colored paving is fairly cool to the touch because it reflects heat.

This glass screen shelters the patio area from cool coastal breezes. It also allows full sunlight to enter. Glass screens are one of the best ways to create a warm microclimate within a cool overall climate.

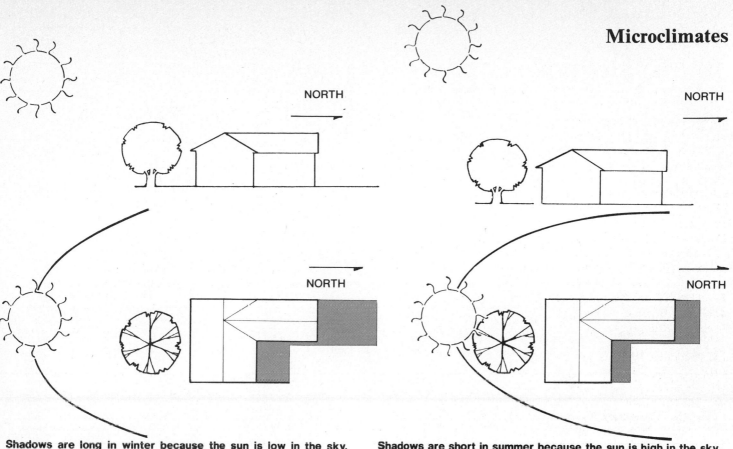

Shadows are long in winter because the sun is low in the sky. Trees, buildings and arbors cast more shade. Overhangs on windows do little to stop solar penetration.

Shadows are short in summer because the sun is high in the sky. Trees, buildings and arbors cast less shade. Overhangs on windows effectively stop solar penetration.

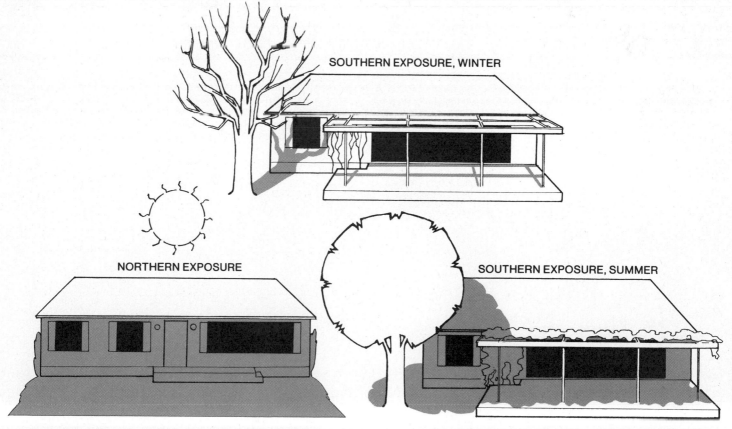

SOUTHERN EXPOSURE, WINTER

NORTHERN EXPOSURE

SOUTHERN EXPOSURE, SUMMER

If you live in a hot climate, locate your patio or deck on the north side of your house. This allows you to exploit the cool microclimate created by the shade of the building. If you live in a cool climate, locate your patio or deck on the sunny, south side of your home. An arbor with a deciduous vine casts summer shade but allows fall and winter sunlight to penetrate when foliage drops. A deciduous tree accomplishes the same goal at less cost.

Microclimates

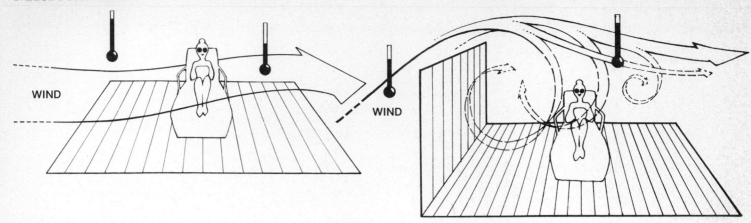

Wind always affects comfort. A patio or deck sheltered from prevailing winds is warmer. Use fences or walls to obtain shelter. A surface exposed to wind seems colder than it actually is because of the wind-chill factor.

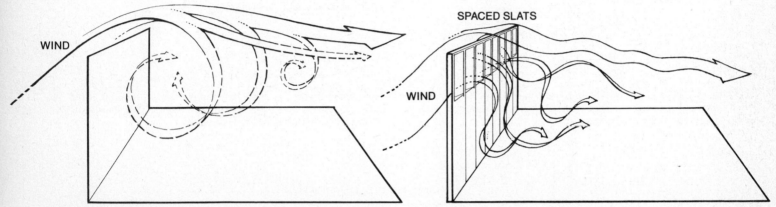

Solid fences and walls make the least-effective wind barriers. A solid barrier has a protected zone behind it approximately equal to its height. When wind hits it, turbulence is created behind the protected zone. As the influence of the barrier weakens, the wind returns to its normal pattern. Greatest shelter is found just behind the barrier within a distance equal to its height.

A fence of slats or lattice that allows some wind penetration creates a semiprotected zone behind it. This zone begins at a distance approximately equal to the fence's height. It extends from this point to behind the barrier a distance of about twice the fence height. These fences protect a greater area than do solid barriers.

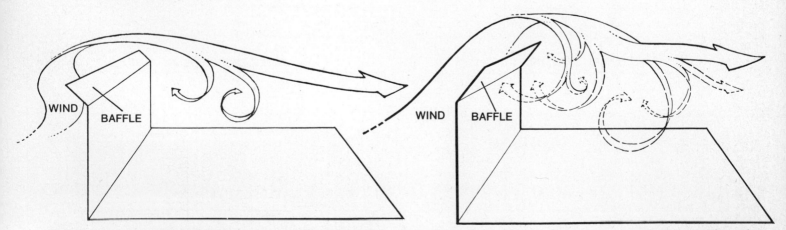

A baffle on top of a fence or wall that angles into the wind creates a zone of effective protection. The zone is approximately equal to twice the barrier's height. A baffle on top of a fence or wall that angles away from the wind creates a warm, protected spot immediately behind the barrier. The protected spot extends slightly more than the wall's height. Wind does not spill over this barrier in the typical pattern. The baffle eases the flow in a specified direction. This reduces turbulence.

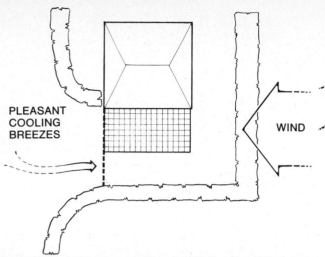

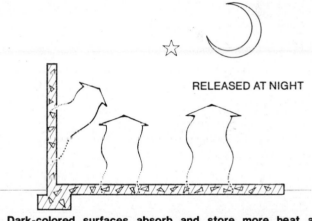

Another way to create a sheltered spot is to sink a patio below ground level. This effect can be simulated by building *berms,* or mounds of earth, around a patio. Planting shrubs on the berms provides additional protection. Avoiding frost is a special problem with patios and decks below ground level. On a still night, cold air flows downhill and collects in low spots. In areas where frost is a problem, allow for *cold-air drainage.* Use shrubs or screens that allow cold air to escape into an area lower than the patio or deck. Deflect cold air around your deck or patio by using hedges, walls or fences. Patios and decks in low areas of the yard also pose special drainage problems. See page 26.

Your house may serve as a windbreak. Windbreaks of hedges or unclipped trees and shrubs are also effective. They can be grown as tall as necessary to extend the protected zone behind them. The length of the protected zone is twice the height of the windbreak. Foliage has a permeable quality that reduces turbulence created as wind slams into the break. Consider shade patterns before you plant a windbreak. Investigate the possibility of opening a breezeway to your outdoor-living area. Cooling breezes are often welcome in summer. Usually they originate from a different direction than storm winds. Try to place windbreaks perpendicular to prevailing winds.

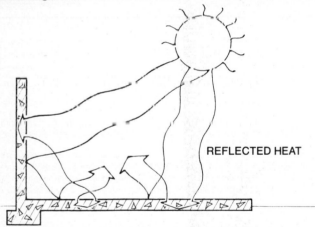

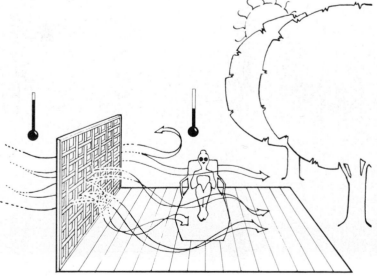

Dark-colored surfaces absorb and store more heat and are warmer. This heat is slowly released during the night. You may want to take advantage of this if you are fond of evening outdoor entertaining. Design a patio of dark-colored masonry.

Surfaces absorb and reflect heat. Paving and masonry walls absorb more solar radiation than wood decks and fences. Light-colored surfaces reflect more heat than dark-colored surfaces and are cooler.

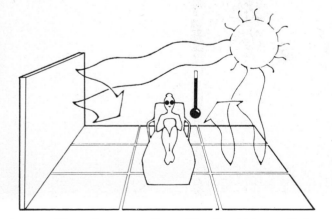

For an area with maximum warmth, position the deck or patio for a southern exposure, and block winds. Use light-colored walls to reflect heat and masonry paving to absorb heat. This maximum-warmth area is ideal for sunbathing.

For an area with maximum coolness, use the north side of buildings or trees and arbors for shade. Masonry that is continually shaded stays cool. If sun strikes the masonry, lower its temperature by sprinkling it with water. Or, you can use a wooden deck which stores little heat.

Design Considerations

Now that you've selected the site and determined how to make it suitable, it's time to design your patio or deck.

First, consider the space you will be working with, as explained below. You'll need to allow space for the activities that will take place. This will determine the size and shape of your patio or deck.

Next, determine how site features will affect the patio or deck structure, as described on page 10. Privacy, grading and drainage, and features such as steps and ramps, screens and overhead structures must be planned to suit the site.

Finally, consider the amenities required to suit your lifestyle. You'll need light if you use the patio or deck at night. Consider such features as water and electrical outlets, barbecue pits, storage areas, planter boxes and seating. See page 29.

HOW MUCH SPACE?

As you plan your patio or deck, assess your needs and the space you have to work with. This will help you determine the size. The simplest way to calculate how much space you need is by use. If you are building a hot-tub deck, allow space for people to sit, hang towels and set down drinking glasses. Add drying-off space and other useful areas as desired.

You can install a patio or deck in a small space as well as a large one. A minimum patio or deck size of 5x6' allows space for two to relax at a small table. To this minimum size, add room for people to walk by without having to move furniture. Allow more space if more or larger furniture is to be used. When designing a small deck or patio, allow just enough space for the activities that will take place there.

When planning space, visualize the furniture layout. Position your furniture and measure the space it requires. If you are buying new furniture, take your measurements to the furniture showroom.

For multipurpose patios and decks, allow approximately 64 square feet for each family member. This square footage can be broken into several pieces for use by children, adolescents and adults.

Scale—This term refers to the size of various patio and deck features in rela-

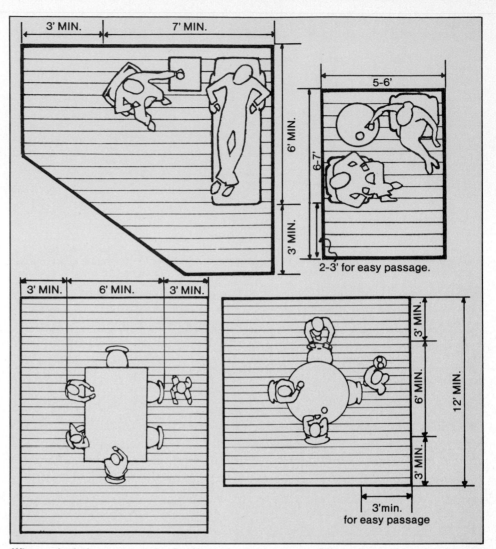

When calculating space, take furniture size into account. Different amounts of space are required, depending on the furniture. You will also need a *zone* that allows people to pass by. Here are some sample zones. If you already have furniture or you know what you will buy, calculate your own zones.

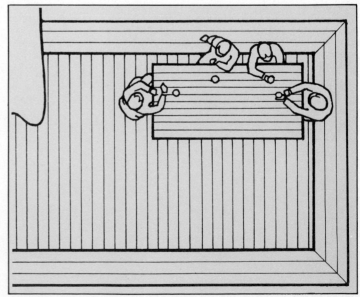

One way to save space is to pull up a table to built-in benches at the patio or deck edge. Benches should be in proportion to the patio or deck. Use wider benches for larger spaces, narrower benches for smaller spaces.

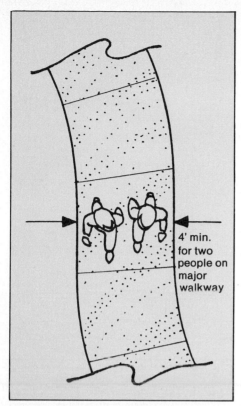

Outdoor-living areas need to be linked by walkways. Wide paths are in scale with large gardens or at house entries. Small paths are appropriate for small gardens or for access to seldom-used service areas.

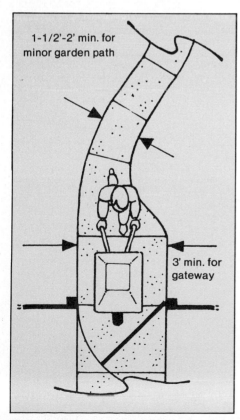

Make gateways wide enough for easy passage. Three feet is a good minimum to allow passage of a wheelbarrow.

One easy way to make spaces compatible is to let the outdoor-living area reflect the size of the adjacent indoor room.

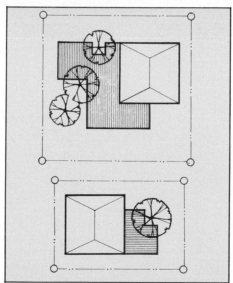

Large patios and decks fit large houses and yards. Small surfaces are suitable in small spaces. One way to calculate space is to let the size of the adjacent indoor room dictate outdoor-living area size.

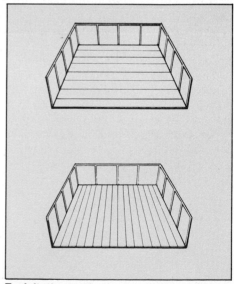

Exploit the eye's sense of proportion. A deck or fence of wide boards makes an area look smaller. One with thin boards makes the area look larger. Placing boards perpendicular to lines of view makes the view seem closer. Placing boards parallel to lines of view makes the view seem farther away. Paving units can be used in similar ways.

tion to the size of surrounding buildings or landscape components. Outdoor scale is not the same as indoor scale. Outside walks, steps and spaces are usually larger than inside ones, simply because the surroundings are more open and spacious.

Keep patio and deck elements in scale with the space you have. Examples of correct scale for patios, decks and walkways appear on pages 18-19.

SHAPES

Once you know where you want your patio or deck, you must design a shape for it. Shown on these pages are some basic design concepts used to determine patio and deck shape. The design you choose must fit the space you have to work with.

You have many choices in shapes, but don't overlook classic squares and rectangles. Pleasing combinations of these shapes add interest to an overall landscape scheme.

Think of patios and decks as *multidimensional surfaces*. Whether on flat ground or a hillside, a multilevel patio or deck creates a visual sense of depth.

When planning patio or deck shape, think about the kinds of trees and shrubbery you want to plant around the area. Surrounding plants can enhance the shape or detract from it. It all depends on the plants you choose and where you locate them.

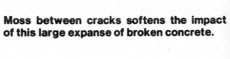
Patio shape is mirrored by plant pockets that visually break the paving and anchor it to the garden.

Moss between cracks softens the impact of this large expanse of broken concrete.

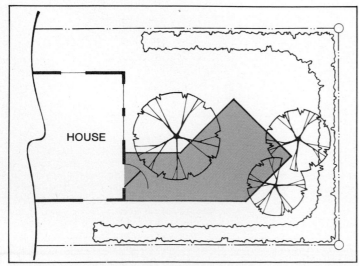
Diagonal lines add interest to a small, boxy yard.

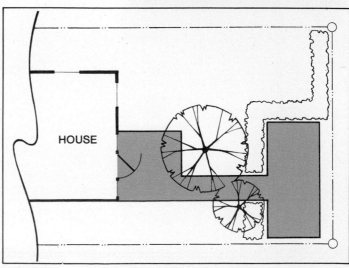
Two platforms connected by a walk create a hidden retreat.

This patio is pleasingly geometric. Its converging lines point to the view and create a sense of perspective.

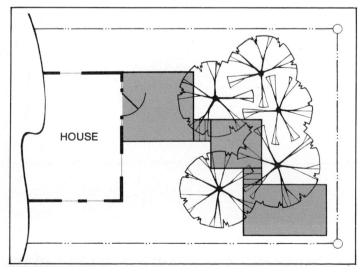

A series of platforms on different levels connected by steps makes a slope usable. Passing through a grove of trees makes the space seem larger and more complex.

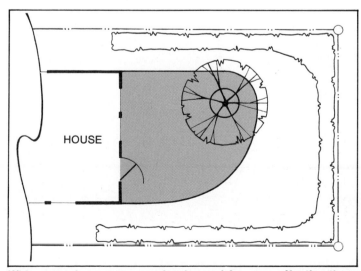

Wide, sweeping curves are pleasing and far more effective than tight squiggles.

PRIVACY POINTERS

Each patio or deck may have several living zones—public, semipublic and private. The more enclosed an area is, the more it becomes an outdoor "room." Here are some suggestions for creating outdoor living areas hidden from the view of neighbors and passers-by.

A solid wall provides maximum privacy and sound-proofing. This inward-looking city patio has a fountain as its focal point. It has no views out. Abundant greenery, simple details in one color and the sound of splashing water make the space seem larger.

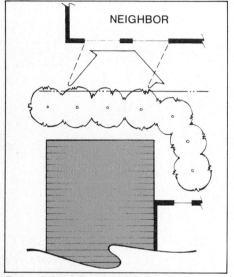

For an informal, natural-looking landscape, screen the lot with fast-growing, unclipped trees and shrubs.

A lattice screen and dense hedge lend privacy to this small condominium deck.

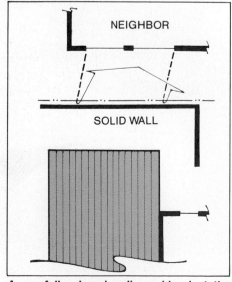

A carefully placed wall provides just the right amount of privacy.

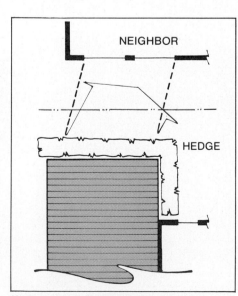

Hedges are living walls. They require more space than a fence, but they provide greenery. For a similar effect in less space, train vines to grow on a wire fence.

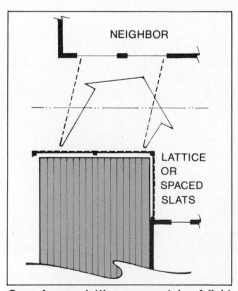

Open fences, lattice or a curtain of light foliage will allow light and breezes to penetrate while providing reasonable privacy.

MINIMIZING MAINTENANCE

You can reduce patio or deck maintenance by planning ahead. Provide areas where leaves and other debris can be swept or hosed off. Built-in benches, tables and other permanent structures should be easy to sweep under. Deck boards should be spaced so dirt does not accumulate in the cracks.

Design the perimeter of the patio or deck to make plant and lawn care easy. If there will be a lawn next to the patio or deck, don't create areas the lawn mower can't reach. Don't block access to garden areas with rails or other structures.

Shown here are a few design tricks that will help speed chores later.

Bench or rail supports are raised off the deck surface to make sweeping easy. Nothing along the edge obstructs a broom.

On a ground-level patio, brooms are not obstructed but lawn mowers are. Ground cover would have been better than grass for easy maintenance around this bench.

Curbing and confining planting to raised beds makes a patio easy to sweep. It helps keep dirt in its place. Hosing this surface is another approach to cleaning. Make provisions for drainage. Curbs act like walls. A catch basin may be necessary for runoff water. See pages 25-26.

GRADING AND DRAINAGE

Grading and drainage are essential, interconnected design considerations. A patio or deck is an exterior surface that must be fitted to your house and to the land. Understanding grading concepts will help you decide which type of structure to build. Plan now so your ideas will work when construction begins.

Paved surfaces usually require *grading*, or soil redistribution, to provide a firm, smooth base. Proper grading ensures correct *drainage*, or directing water off the surface, away from the house. Grading is done by a *cut-and-fill* process, removing soil from high spots and using it to fill in low ones.

Decks require firm footing in soil, but major grading is rarely necessary. Decks are appropriate on sloping sites or around trees where it is not desirable to disturb existing contours, such as for a patio.

Grading or filling more than 3 to 6 inches of soil within a tree's *dripline* causes serious root damage. The tree may die as a result. See drawing on page 26. Trees in cut-and-fill areas must be protected with a *dry well*—a hole filled with gravel. Consult a tree specialist for construction details.

The basic principle of surface drainage is simple: Water runs downhill. Design the surface to create low spots and direct the water into them. A patio should never be flat. It should slope into planting areas or to a collection point, usually a *catch basin*. If designed properly, the patio slope should be almost imperceptible to the eye.

Subsurface drainage is the flow of water under the soil. Water enters a catch basin. The water then flows from the basin through a pipe to a lower collection point. Water that collects behind a retaining wall must be drained away in an underground *drainpipe* or through *weep holes* drilled in the wall. Otherwise the wall may eventually collapse. For more information on drainpipes, see facing page.

Gravel and sand are subsurface materials that drain water more rapidly than clay soil. Subsurface drainage is expensive, and there is little room for error. If you need a subsurface drainage system, check local codes and determine where you can dispose of the water.

It is illegal to direct water onto your neighbor's property if that is not the natural drainage pattern. You may direct water into a natural drainage *swale*, or low area, or into a creek that flows through your property to a neighbor's. You may not dig a ditch to the edge of your property and send all the water from your patio and downspouts into the ditch. Drain water into lawn or planting areas, natural drainage swales, dry wells, storm sewers or a street gutter that empties into a storm sewer.

Topsoil is the rich, dark surface layer that supports plant life. Before any grading work is done, remove the topsoil and pile it in an out-of-the-way place. This exposes the *subsoil,* or *subgrade,* underneath.

Moving the subsoil to conform generally to the desired surface contour is called *rough grading.* This can be done with a shovel, or for larger areas, with an earth mover. Replacing the topsoil and raking the surface to final contour is called *finish grading.* The job is completed by thoroughly watering the soil to settle it. Don't replace the topsoil if paving will cover the area. Save this soil for another part of the garden.

Always water and thoroughly compact the subsoil with a *tamper.* This tool can be rented at tool rental outlets. See page 103. Tamping the soil will provide a firm base before the patio is installed. You may have to remove some of the subsoil to accommodate a layer of gravel or sand beneath the patio. In some areas of the yard, loose gravel itself makes a good patio surface. Water percolates through it to the soil beneath, so there is less runoff.

Deck platforms step down a slope. This eliminates the need for soil cut and fill, which endangers tree roots.

Fine old trees are the focal point for this patio. Broad, attractive steps lead from porch down to brick-on-sand paving. This paving permits water and air to reach tree roots.

There are two specialized surface-drainage situations to consider—roof decks and watertight below-deck storage areas. Wooden decks or masonry surfaces on rooftops must not interfere with existing roof-drainage patterns. Remember to drain runoff into downspouts as you design new roof structures. For decks built on existing roofs, joists laid perpendicular to surface flow can be notched to permit runoff to pass through.

The area below a raised deck can be useful for storage. The deck or paving must form a watertight roof and provisions should be made for proper drainage.

Grading to obtain a natural look and proper drainage is an art. For sites with multiple slopes or designs that ignore natural contours, hire a licensed landscape architect or engineer. They will make a grading plan. This plan shows spot elevations at all critical points. It also shows existing and proposed contour lines. Such a plan is useful if a contractor is to do your grading.

DRAINAGE SYSTEM DESIGN PROCESS

1. Determine existing runoff patterns. Find high points, low points and natural swales.
2. Decide where you want to direct the water. You may choose to have water percolate into the soil or flow into a natural drainage swale. You may want it to flow to a collection point that ties into an off-site drainage system.
3. Determine the obvious places that will collect water—walled or sunken patios, narrow pathways between two buildings, areas around a pool, spa or hot tub.
4. Decide whether surface or subsurface drainage best handles the problem.
5. Determine where the water will be discharged. For surface drainage, plan where it will leave the property or will percolate into the ground. For subsurface drainage, plan where the drainpipe will go.
6. Slope the ground surface or the underground drainpipe from area to be drained down to the discharge area. Maintain the minimum recommended slopes for surfaces and pipes. See "The Meaning of Grade Percentages" on page 26.

A narrow path can usually be

A catch basin collects surface water and delivers it to a subsurface drainage system of pipes.

A tiny rivulet in this old New Orleans patio collects and carries away surface water. The surface slopes to the rivulet. The rivulet slopes out of the enclosed courtyard. This is an ancient type of drainage system. It eliminates the need for catch basins.

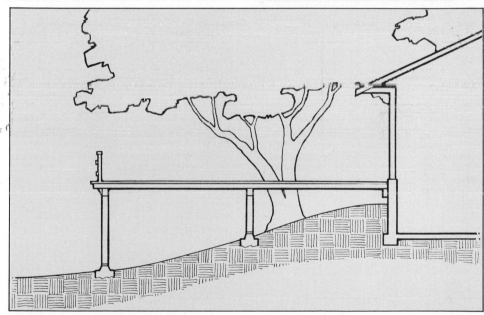

Preserve an existing natural slope by building a raised deck. No cut or fill is necessary. Grading is only required if soil slopes back toward the house by builder's error. Use space under raised decks for storage.

crowned, or mounded, so water drains from the center to either side. It can also be *pitched,* or angled, so water drains across it. Don't forget to pitch steps slightly. Walled patios need catch basins. See drawing on page 26.

Water discharges from a drainpipe at its *outfall* and enters a gutter, storm sewer, ditch, creek or dry well. Obviously, the pipe outfall must be lower than the area to be drained.

The drainpipe must slope a minimum of 1%. Otherwise silt will settle from the flowing water and clog the pipe. Don't join drainpipes at right angles. Angle two lines together so they form a "V" that points downhill. Flexible plastic pipe works well in this situation.

Perforated drainpipe is installed with holes pointing *down.* This positioning permits water to rise into the line or, conversely, to percolate down into a gravel drainage area. It also prevents dirt from entering and clogging the line.

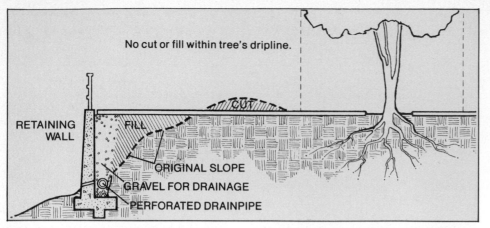

No cut or fill within tree's dripline.

RETAINING WALL

FILL

ORIGINAL SLOPE

GRAVEL FOR DRAINAGE

PERFORATED DRAINPIPE

Cut and fill for a terrace or patio. This process creates a level area by cutting down a high spot of earth to fill a low spot. Always remove and save the topsoil first. If you can, balance cut with fill. If you have excess dirt, you may be able to use it in some other part of the yard. If you need more dirt for fill, you can take it from another part of your property or buy it from a sand-and-gravel dealer.

Never cut and fill more than a few inches within the dripline of trees you wish to preserve. Retaining walls are necessary to create level areas on sloping sites. Provide drainage behind retaining walls. Most codes require retaining walls over 3 feet high be designed and installed by a qualified landscape designer or a soil engineer.

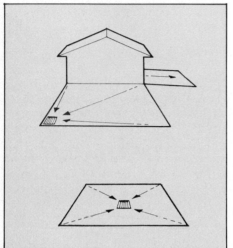

Patios must drain away from the house. The top patio is sloped in one direction away from the house foundation. This is an example of surface drainage. The bottom patio slopes inward from the edges to a catch basin in the center. Water in the catch basin flows into an underground drainpipe that slopes away from the patio. This is an example of *subsurface drainage*. The catch basin in the top example could drain directly into an adjacent planting area if it is lower than the patio surface.

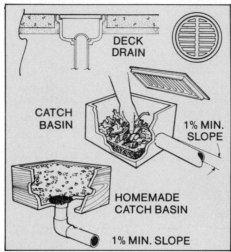

DECK DRAIN

CATCH BASIN

1% MIN. SLOPE

HOMEMADE CATCH BASIN

1% MIN. SLOPE

Your patio design may call for center drainage. For small areas, use a *deck drain*, shown at top. For larger areas, use a catch basin. The prefabricated catch basin, center, has a removable grate for cleaning out leaves and debris. You can make your own catch basin as shown at bottom. Use pressure-treated lumber for the basin box. Install a drainpipe at the bottom. Cover the drain opening with a screen. Fill the box with gravel. No grate is necessary.

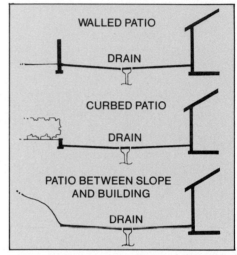

WALLED PATIO

DRAIN

CURBED PATIO

DRAIN

PATIO BETWEEN SLOPE AND BUILDING

DRAIN

When designing your patio, avoid using catch basins. They increase costs and are not foolproof. But they are essential when water has no way out of a walled, curbed or sunken patio, as shown here. Catch basins are necessary where earth forms—slopes or mounds—prevent runoff. For an ancient alternative to catch basins, see the photo on page 25, top right.

THE MEANING OF GRADE PERCENTAGES

Slope is the ratio of rise or fall of a surface or drainpipe to its *run,* or length. This ratio is expressed as a percentage. For instance, a slope of 1/2% (0.005), equals a 1/2-foot (6-inch) rise or fall in 100 feet of run. This converts to roughly 1/16-inch rise or fall per foot of run, or 1-inch rise or fall in 16 feet of run. This is the absolute minimum slope for smooth concrete, terrazzo, marble and glazed tile.

A 1% slope (0.01), equals a 1-foot rise or fall in 100 feet, or roughly 1/8 inch in 1 foot, or 1 inch in 8 feet. This is the minimum slope for pipes, brick, quarry tile, semismooth concrete or exposed aggregate.

A 1-1/2% slope (0.015), equals an 18-inch rise or fall in 100 feet, or

roughly 3/16 inch in 1 foot, or 1 inch in 5 feet 4 inches. This is the minimum slope for very rough exposed aggregate, stones set into concrete, flagstone or asphalt.

A 2% slope (0.02) equals a 2-foot rise or fall in 100 feet, or roughly 1/4 inch in 1 foot, or 1 inch in 4 feet. This is the minimum slope for rough-surfaced materials, including loose gravel, tanbark and grass. The maximum slope for these materials, except lawns, is 3%. A 2% slope is often recommended for surfaces to be installed by unskilled homeowners. This percentage leaves a margin for slight errors in construction. It is also the maximum slope for most paved surfaces. Slopes of more than 2% become noticeable,

an undesirable condition for patios. Paths and ramps may be sloped up to 8%. There is no maximum on pipe slopes. A 2% slope will drain more effectively than a 1% slope.

A 3% slope (0.03) equals a 3-foot rise or fall in 100 feet, or roughly 3/8 inch in 1 foot, or 1 inch in 2 feet 8-1/2 inches. This slope becomes obvious in relation to level areas. It is the minimum slope for ground-cover areas.

An 8% slope (0.08) equals an 8-foot rise or fall in 100 feet, or roughly 1 inch in 1 foot. This is the maximum slope for ramps accessible to the handicapped. It is difficult to push wheelbarrows up steeper ramps or to mow lawns on steeper slopes.

STEPS AND RAMPS

Use steps or ramps to make an attractive link between two levels. There are many possible step designs and construction methods. Choose materials that harmonize with the patio, deck, house or landscape.

Steps are composed of *treads,* the walking surface, and vertical *risers.* The most important step design criteria is safety. Abrupt and unexpected changes in tread width or riser height are hazardous, so treads and risers within a flight, or run of stairs, should always be consistent. A single step is also a hazard because the eye tends to miss such a slight change of grade. The recommended minimum number of steps is three. For areas with slight slopes, a few small decks or platforms at different levels can take the place of steps. Ramps are often more useful and inexpensive because they ease the passage of wheeled vehicles and devices. Illuminate steps and ramps for safety at night. Step and ramp construction is often governed by local codes. Check with the building department for requirements.

Generous scale of steps and ramps

Well-built brick steps are necessary for the steep grade change created by the retaining wall encircling this sunken patio.

Simple slabs of local stone make a natural transition from patio to slightly elevated lawn.

outdoors is more pleasing than the narrow, steep-staircase effect used in indoor spaces. Broad, low steps can double as seating or plant-display platforms. Steps 5 feet wide allow two people to walk comfortably side-by-side. Wider steps are suitable for grand-scale effects. Break long flights of steps into several flights connected by landings. Keep the tread-riser relationship consistent between flights.

A 6-inch riser and a 12-inch tread is perhaps the most useful outdoor step

relationship. These dimensions are well suited to outdoor scale. It is easy to calculate the necessary number of steps. Just divide each 1-foot elevation change into two 6x12" steps.

When figuring tread-riser combinations, use this old formula: Two risers plus one tread should equal 23 to 25 inches. This is the basis for standard outdoor tread-riser combinations. The wider the tread, the lower the riser. See chart on page 156.

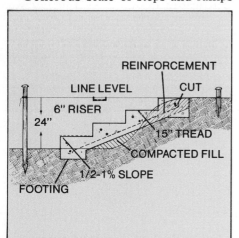

To determine change in elevation, place a short stake at the top of the slope and a tall stake at the bottom. Connect string between the stakes and level it with a line level. Measure the length of the string, then measure up the tall stake to determine the grade change. With these measurements, you can figure the tread-riser relationship of the steps and how many steps you'll need.

Cut and fill soil as necessary to accommodate construction. To provide a firm base for the steps, thoroughly tamp the ground to compact it. A flight of concrete steps requires metal reinforcing rods. It also needs a concrete footing below the soil and a slight pitch of 1/2 to 1% on each tread to drain off standing water. Concrete steps can be surfaced with flagstone or rough-surfaced patio tile.

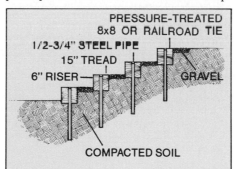

Railroad ties or pressure-treated 8x8s are anchored with countersunk pipes to form the risers for these steps. Tread area can be covered with gravel, bricks or other material. A second tie behind each riser is another tread option.

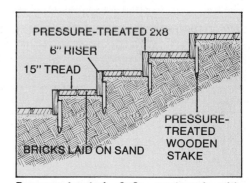

Pressure-treated 2x8s anchored with stakes are attractive in combination with bricks on a sand bed. For wood-step construction details, see pages 155-156.

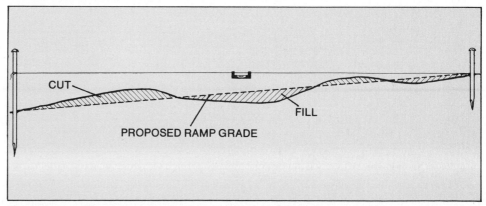

Use ramps at 8% maximum grade where space is not limited. Cut and fill to even surface irregularities. Thoroughly tamp soil before laying paving, decking or gravel.

ONE DESIGN SOLUTION

Here is one of many possible designs for the fictional property we followed through base map, site analysis and bubble plan on pages 12-13.

Space for activities, patio and deck shapes, privacy, maintenance, grading and drainage have all been considered. The design now includes actual dimensions, materials and design details. Plants, screens, storage, outdoor-living areas and drainage patterns are established.

The front driveway is screened by trees that cast shadows to break up the expanse of hot asphalt. A landing pad and an enlarged walk lead to an entry court. Short sections of fence in a baffle arrangement ensure privacy and eliminate the need for gates.

Fencing is within code setback limits. A bench surrounds the existing tree that is the courtyard's focal point. A new gate screens the old walkway, which has been extended to the back patio. Hedges along the property line provide privacy. A gap in the hedge allows a view of the creek outside the property. A second hedge at the house corner conceals most of the yard from outside view.

Another tree has been added to make a line and create a new sense of space in the back yard. The tree also

helps screen the hot-tub deck and shelter the main deck. A short hedge opposite the tree cutout in the main deck divides the deck into two smaller areas.

The barbecue-and-storage unit on wheels rolls where needed or parks out of sight on a paved pad adjacent to the house. A bench around the tree provides seating or a spot for sunning. The larger deck is used for entertaining. A ramp leads to the hot-tub deck screened by a lattice fence and shrubbery.

Hot-tub equipment is concealed under the deck. Overflow water enters a catch basin. This separate retreat is completely sheltered from wind. A gravel service area for storage and potting is created by screening off a section between the fence and patio. Shrubs help break the wind. A gravel path leads around the house to a tiny kitchen patio. A lattice screen lends an open, airy feeling to the area.

Except for one catch basin next to the hot-tub deck, all runoff water drains on the ground from the house to the lot's lowest corner.

This finished site plan is rendered on the base map. By using 100%-rag tracing paper as suggested on page 12, you can make blueprints of your design. Make several blueprints.

Submit one for a building permit. Keep one for your records. Use one to mark up as you estimate materials. Give copies to your construction helpers or to the contractors who will bid for the job.

CATCHING MISTAKES

It is difficult to make changes when construction is under way. When work is complete, you may find that the sizes are all wrong. Laying out your proposed patio or deck design with stakes and string will help prevent errors. You can then visualize spaces and the relationship of various elements.

String up bed sheets to determine the shape of storage units and the best height for privacy screens and arbors. Walk through the proposed design. Make note of use areas, traffic circulation patterns, lighting and electrical outlet locations. Stake out built-in benches and planters to determine proportions. Check everything for scale.

Finally, sketch in your proposed design, making adjustments on tracing paper overlays.

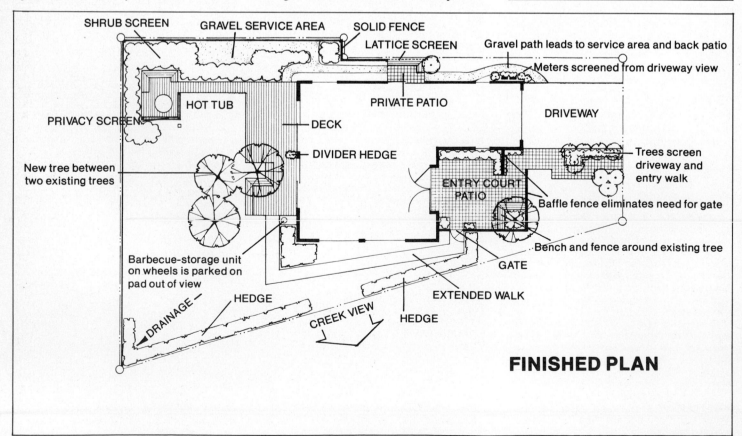

SHRUB SCREEN GRAVEL SERVICE AREA SOLID FENCE
LATTICE SCREEN
Gravel path leads to service area and back patio
Meters screened from driveway view
HOT TUB
PRIVATE PATIO
DRIVEWAY
PRIVACY SCREEN
DECK
New tree between two existing trees
DIVIDER HEDGE
Trees screen driveway and entry walk
ENTRY COURT PATIO
Baffle fence eliminates need for gate
Barbecue-storage unit on wheels is parked on pad out of view
Bench and fence around existing tree
DRAINAGE
HEDGE
GATE
EXTENDED WALK
CREEK VIEW
HEDGE

FINISHED PLAN

DON'T FORGET AMENITIES

Remember to plan amenities that will make your outdoor-living area more useful and enjoyable. The most important of these is lighting. Patios and decks adjacent to the house can be illuminated with simple floodlights on the house wall. Subtler lighting will make the outdoor space more inviting and help create an intimate atmosphere for parties. Use special fixtures or lanterns hung from arbors or trees. Trees or shrubs lighted by below-ground-level lights become sculptural focal points at night. This lighting effect is called *uplighting*.

Always light steps, ramps and changes in deck levels. Also light gates and entries. Install one or two outdoor electrical outlets when you wire for lights. Put the outlets in convenient locations for operating electric power tools and devices.

One of the most important but overlooked patio or deck additions is a simple *hose bib*. It can be used for quick cleanups, cooling hot paving and watering plants. Hose bibs are easy to hide. Include space for the hose. Hidden storage is an important design consideration. See the suggestions on pages 34 and 84.

Drinking fountains are always appreciated near hot tubs, tennis courts and children's play areas. You can buy inexpensive fountain fixtures that attach to hose bibs. Showers on beach decks help keep sand outdoors.

Firepits and special heaters help extend the useful season of your outdoor-living area. Built-in barbecues are popular patio and deck additions. Concealed garbage cans are useful near outdoor-eating areas.

Consider adding planter boxes and hanging baskets to your design. A built-in drip-irrigation system will make decorative planters practically carefree. Planter boxes can be more than decorative. Substitute them for rails on raised patios or decks or use them as benches. Remember to include the weight of planter boxes when you design the structural support for a deck. Snow also has an influence on structural design. Check with your local building department for code requirements.

Here is a method for snow disposal for decks adjacent to house entries. Set a grate into the deck. Stamp your feet and snow drops off boots through the grate.

Lighting is essential at steps and entries.

These simple light fixtures complement deck design and make it inviting and useful at night. The planter box serves as a rail.

Don't forget the ever-useful hose bib. Duck-board decking eliminates muddy drainage areas.

Narrow storage area was built into the back wall of this partially enclosed deck.

Before You Build

When you have completed the plan for your patio or deck, you must decide who will build it. You have the option of doing all of the work, some of it or none of it.

You can save money if you build part or all of the patio or deck yourself. Hiring a contractor to do the entire job costs more. Despite the extra expense, there are good reasons for hiring an expert.

Contractors usually work faster than most weekend craftsmen. They have specialized equipment and employ experienced laborers. This combination generally produces good results. When you hire a reputable contractor, you have confidence the work will be done properly. If problems occur, you can usually trust the contractor to make necessary corrections.

Patios and decks that involve an awkward site or a peculiar design may require professional help. Decks more than 6 feet above the ground, complicated drainage systems or potentially unstable slopes should be handled by a qualified expert. Codes require that retaining walls more than 3 feet high be professionally engineered. Decks that must be leakproof or ones that will carry heavy loads require careful attention. So do patios and decks built over water.

CONTRACTORS

The type of contractor you hire depends on the project. Landscape contractors are the most versatile. They'll build masonry patios, wooden decks, fences, walls and other garden structures. They also do planting. Most landscape contractors are set up to contract all of the work indicated on a landscape architect's plan. They often subcontract specialty work, such as pouring a concrete patio.

You may want to hire a specialty contractor for a certain phase of the work and do the rest of it yourself. It may be less expensive to use a specialty contractor for the whole job.

Masonry contractors build masonry patios and walls. Some general building contractors and fence contractors build wooden decks. If you're building the patio or deck, consider hiring a licensed electrician to install lights or outdoor receptacles.

How do you choose a reputable contractor? First, check his references, then ask to see examples of his work.

Check with the Better Business Bureau and the State Contractor's License Board to see if any complaints have been filed against the contractor's company. Make sure the contractor is insured for property damage, public liability and workmen's compensation.

When checking examples of the contractor's work, ask the homeowners if they were satisfied with the working arrangements and the completed project.

Before choosing a contractor, get at least three competitive bids. A complete set of plans will help the contractor make a more accurate estimate. The bid should itemize all materials and labor.

Writing A Contract—If you do hire a contractor to do the work, protect yourself with a written contract. It should specify grades of lumber, kinds of preservative and finishes, types of hardware and other important construction details. It will also include payment and work schedules.

A good contract will have a payment schedule that calls for progress payments as each phase of work is completed. The final payment should be due after you've had a reasonable amount of time to inspect the finished project. The contract should clearly state what services the contractor has promised to provide. Don't accept any verbal promises. Get them in writing.

DOING IT YOURSELF

You may have the necessary tools, skills and confidence to do careful work. If you decide to do the work yourself, you'll have to get the necessary building permits. You'll have to call for building inspections at various stages of construction. Your city or county building department can advise you on permit and inspection procedures in your area.

PERMITS

Most cities and counties require a building permit for constructing patios and decks. If you're doing any electrical work, you'll also need an electrical permit. Your plans must comply with local codes before you can get the permits. Neat, detailed plans will help speed the process of getting your permit.

If regulations prohibit your best idea, apply for a *zoning variance*. The variance is written permission from city or county officials to bypass certain codes. Letters from neighbors stating that your plans are not objectionable will sway officials in your favor. There may be a small fee for a zoning variance.

A wooden deck is a good do-it-yourself project for those with average carpentry skills. Most decks can be built using ordinary carpentry tools.

Ideas From The Professionals

The following interviews with professional landscape architects and designers serve as a source of ideas for integrating patios and decks into your landscape. Some of these ideas can be incorporated into your plans. The interviews also provide insight into how professionals approach design problems and meet the needs of their clients. This will be of help if you decide to hire a landscape architect or designer.

STEPPING DOWN TO A BEAUTIFUL VIEW

Robert A. Dean of Wimmer, Yamada & Associates in San Diego, California, describes how his firm designed a deck with easy access to the top floor of a two-story house:

"Our clients had a large, unused area next to their home. They asked us to design a deck that would be comfortable for outdoor living in the evening and still function as a recreation area during the day.

"Upon analyzing the site, we saw that most of their daily activity took place on the second floor of the house. The kitchen and living room were located there. The bedrooms were on the bottom floor.

"We decided to come off the second-floor deck with a stairway that would blend with the existing architecture and allow people to look down and view the main deck.

"The lower deck is done in hexagonal patterns. It's large enough for sunning or entertaining and for the spa. It's raised just high enough to enjoy a beautiful view of the Pacific Ocean but it's still subject to frequent winds. To solve this problem we installed glass screens on the west side of the deck. The panels are louvered so they can be opened and closed as need be."

Glass screen with louvered panels blocks wind while preserving an ocean view.

This deck is designed to complement the house design. Stairs provide a smooth visual transition and comfortable ascent to the second story where most family activity takes place.

DESIGNING AN ENTERTAINING TERRACE

Leon Goldberg is one of the partners of Goldberg and Rodler, Inc., of Huntington, Long Island, New York. He tells how his patio design improved indoor and outdoor living conditions:

"Our clients had given the yard a great deal of thought and knew what they wanted. All they needed was professional assistance to execute their plan.

"They had done some construction within the house. This included a glass wall in back that opened onto a raised patio. The patio was uncomfortably hot and sunny in summer. The clients felt it did not complement the nice work they had done indoors. They wanted something that looked good from indoors. It also had to be a comfortable place for outdoor activity, which is a big part of their summer life.

"We added an overhead for shade and covered the existing cement patio with bricks. The overhead also cut the glare that entered the house through the windows. It enclosed the patio and created an outdoor room. The overhead framed the view of the large lawn and wooded area behind the house.

"The patio is loaded with flowering plants in containers. A vine partially covers the overhead for additional shade. All in all, the patio is an extremely attractive extension of the house."

Bright furniture and containers add color and comfort to outdoor living.

The brick patio is terraced with pressure-treated beams to bring it up to the floor level of the house. An arbor frames the patio, creating an outdoor room. It also provides shade, indoors and out, and serves as a visual connection between house and yard.

CREATING AN OUTDOOR ROOM

Rudi Harbauer works for Atlantic Nursery Landscaping, Freeport, New York. He describes several design solutions for a small, odd-size lot:

"The client's yard presented several challenges. First, it was pie-shaped and very small. It was only 8 feet wide at the narrow end. Second, it was bordered by a two-story house which was extremely overpowering.

"They wanted an outdoor area for entertaining and eating. The back door to their house was 4 feet above ground level. We built a narrow boardwalk along the side of the house. This emptied into the wide end of the yard. This way we didn't eat up any of the main back yard with a series of steps. A barbecue was placed at the end of the boardwalk near the door to the house. Now the owners won't have to step all the way down into the garden to cook.

"The patio was surfaced with brick. We created a Japanese garden setting at the narrow end using rocks, gravel and dwarf conifers. To give the area more depth, we dropped the fence level from 6 feet to 4 feet, half way down the yard. We covered the lower portion of the fence with flowering shrubs.

"Because the house was so tall and the average width of the yard only 12 feet, a bigger fence would have given the yard a claustrophobic feeling. The 4-foot fence gave privacy without closing in the yard.

"To further buffer the height of the house, we added an open-air, U-shaped trellis. It's perfect for hanging baskets and contributes to the outdoor-room feeling."

A narrow deck walkway, an arbor and a brick patio turn the smallest yard into an intimate outdoor room.

FILLING A NARROW SPACE

Landscape architect James Loper lives with his wife and two children in Louisville, Kentucky. He tells how he solved an urban-living problem by designing and building a deck:

"With two young children who can't always be accompanied to a park, we needed an outdoor play area. This presented a problem. Our back yard measured only 16x45'. It was bordered on the rear by a 4-1/2-foot-high concrete retaining wall. The wall separated the yard from a service alley. My wife Susan was afraid the children would fall into the alley.

"We observed where sunlight fell and where shade was cast. We determined the direction of summer breezes and where our best and worst views were. Then we focused on our needs for an outdoor living and entertaining space. We considered our children's need for a safe place to play. I knew our small yard offered opportunities a large yard didn't. For one thing, we would be able to develop and landscape the entire site.

"The first step was to measure our site and prepare a base map. As I worked with pencil and sketch paper, the concept of a large wooden deck with a 6-foot privacy screen evolved. This deck would allow us to enclose the view from our kitchen. It would also allow us to step out onto the deck at the same floor level as the kitchen. The space under the deck would provide secure storage space for lawnmowers, potting soil, wheelbarrows and bicycles. We would still have access to the basement door.

"The northern end of the yard receives the drying warmth of the sun early each morning. This made it the best location for the playground. The central area would be retained for circulation and be developed as a bricked courtyard with wrought-iron furniture. The side yard, which was 18 inches higher than the rear yard, was wrapped around the edge of the house by means of a retaining wall. This created three different levels within the yard. It also added to the interest one feels when looking down on the courtyard and playground and up to the deck. It shortened the climb to the deck for entering visitors. The basic plan met our needs and the site's constraints.

"My wife and I began to wonder how our neighbors would feel about such a development adjacent to their

Before: Narrow back yard provided little outdoor-living space.

After: Yard is divided into separate activity spaces. Area beneath raised deck is used for storage. The lower brick patio provides a place for outdoor cooking.

34 Planning Your Patio Or Deck

yards. We took our plans to them and discussed what we were considering. They were as enthusiastic as we were. They, too, wanted to do something with their yards in the next few years. They were interested in seeing how such a development would turn out.

"With plans in hand, we went to request a building permit. Our plans violated both side and rear-yard setbacks. This problem is common when improvements to older homes are proposed in neighborhoods developed before current zoning ordinances. An application was made for a zoning variance at the Board of Zoning Adjustments. A visit to your local Board of Zoning Adjustments to discuss your plans is very beneficial. It helps you determine what concerns the board will be considering in evaluating your variance request.

Plenty of seating for eating and relaxing is essential for comfortable outdoor living. Making pear cider is favorite fall activity for the owners.

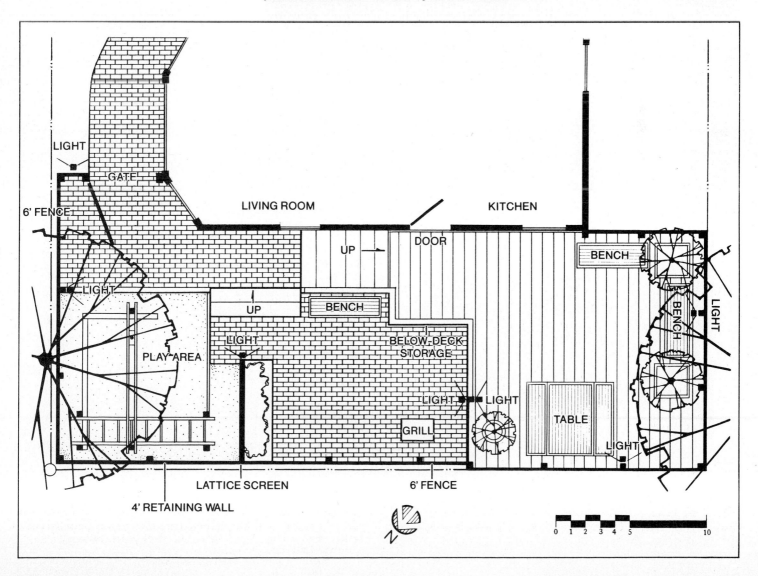

LIGHT

GATE

6' FENCE

LIGHT

LIVING ROOM

KITCHEN

UP

DOOR

BENCH

UP

LIGHT

BENCH

PLAY AREA

LIGHT

BELOW-DECK STORAGE

BENCH

LIGHT

LIGHT

LIGHT

TABLE

LIGHT

GRILL

LATTICE SCREEN

6' FENCE

4' RETAINING WALL

0 1 2 3 4 5   10

"A public meeting was to be held three weeks later concerning our variance request. A bright orange sign was placed in our front yard advertising that fact. I presented our plans at the meeting and the board voted to grant us a variance.

"Here's a tip for those who might become involved in a similar situation. A short letter from your neighbors to the board, expressing their agreement with your plan, helps.

"Armed with variance and building permit, Susan and I went to work in our spare time. We began in the fall. Before winter, we had the deck framed, the steps in, the retaining wall built and perimeter fence posts mounted. We had also purchased and stacked 1,500 used solid bricks. The project was completed the following spring.

"For night use, we wanted a soft-lighted atmosphere. Lights were designed and built to cast downward lighting that would not shine in our eyes. Seven 40-watt lights were used. The bricks were laid in a running-bond pattern on a compacted bed of sand.

"We enjoyed doing our own planning and landscaping. My education and work experience as a landscape architect was very helpful. Other homeowners lacking this background can be assured of getting a safe, functional product by hiring professional planning help."

Children's play area is adjacent to lower patio. Shredded bark provides a soft landing surface under the swing.

MAKING USE OF UNUSABLE SPACE

Gary Martin is a landscape architect working for Landscape Innovations, Melville, Long Island, New York. He describes a project in which he used redwood decks and brick patios to gain space on a small, sloping lot near the waterfront:

"The challenge of this property was its size and shape. It was a long narrow lot with only 24 feet between the house and the waterfront bulkhead. In that 24 feet there was a 7-foot drop in elevation. This drop was too steep to allow a lawn.

"Our objective was to gain usable space in an area that was relatively 'unusable'. We accomplished this by creating a series of decks and patios on various levels. We met the owners' needs for easy access to the water in this way. We also gave them room for entertaining and outdoor living.

"The first deck of the series started at the back door of the house. It ties right in with the kitchen and den. The door actually opens into the kitchen, but den and kitchen are side by side. The interior living space connects with the exterior living space very well.

"The decks were made of redwood. Redwood needs little maintenance. Its weathered look blends nicely with the split cedar shingles on the house.

"Drainage problems were minor. We put in drain pipes to carry water away from a few low spots and from gutters on the roof.

"Although you don't always want them, it's usually mandatory to use railings with this type of raised deck. In some instances, we combined the railings with benches. In others we used rope inserts between posts. These railings offer safety but are not as heavy-textured as wood railings would be. We also used redwood planters to break up the expanse of decking. They blend well with the deck and house and are usually filled with annual flowers."

"The first deck comes off the house at a 45-degree angle. Most of the other lines in the design repeat that angle. By doing this we made the small yard feel bigger.

Multilevel decks and stairs span the steep slope behind the house.

"We planted a couple of honey locusts for shade. Even with the light breeze that comes off the water, it gets pretty hot without shade. Otherwise, all the plantings were low maintenance as the client requested."

Raised redwood deck and brick patio create an outdoor living area where no flat space existed. Small deciduous tree will eventually provide cooling shade.

DESIGNING FOR COMFORT AND FRESH FRUIT

Photographer-designer Michael Landis created an outdoor area that serves a dual purpose for his client in Davis, California. It provides a comfortable place for outdoor living in a hot climate and it produces food. He explains:

"As in many parts of the Western United States, the climate in Davis, California, is very hot in summer. Temperatures of over 105F are not uncommon. Yet it is also a climate that is ideal for growing many types of fresh fruit.

"The client wanted a comfortable outdoor area with easy access to the house and the pool. We built a deck with an arbor for shade against a hot, south-facing wall. The fact that the owners are Greek gave us some clues for the landscaping. We took a Mediterranean theme and used grapes over the arbor for additional shade. A variety of citrus was used around the yard and pool. Olive trees were planted in front of the house.

"To create a smooth transition from the arbor to the rest of the yard, we contoured the deck out around the pool. The pool was trimmed with native rock that flowed into a retaining wall in the rear of the yard. Deck boards were custom cut to fit neatly against the rock.

"We used mounds of soil to add interest and break up the squareness of the lot. A screen of redwoods blocks wind and provides privacy."

Deck boards are custom-cut to follow the shape of rocks that line the pool. Rocks and wood blend naturally in a beautiful use of materials.

Swimmers move easily from deck to pool. Along the fence to the left are espaliered apples and genetic-dwarf peaches.

What was once a hot, south-facing wall is now a comfortable place for outdoor living. Low-level diving board blends nicely into the deck.

Grape-covered arbor yields cool shade and fresh fruit. Deck follows the contour of the pool for a smooth transition into the landscape.

Fresh strawberries and peaches are within easy reach of swimmers. Small lawn area at far side of the pool makes an isolated island for sunbathing.

BREAKING UP A FLAT SPACE

Gary Blum of Ireland's Landscaping, East Norwich, New York, speaks of two of his designs that involved patios or decks:

"The client's property was very flat, sandy and right on the water. They wanted something unusual. Their landscaping had to blend with the surroundings and meet the needs of active boaters, which they are. We gave them a seaside look by using multilevel decks.

"The property was totally flat. We brought in over 400 cubic yards of topsoil to sculpt berms on the lawn. This made it appear less flat. The multilevel decking, the raised planters and the earth sculptures created interest and broke up the flatness. Even the pool is sitting on a plateau. We had to raise it because of the high water table.

"Wind is usually a problem in seaside areas. This site was no exception. We used Japanese black pine to create sheltered spots around the deck adjacent to the house. Some of the berms also provide protection from wind. Even the house was designed so the low side was into the prevailing wind.

"Most of the plants we used lend themselves to seaside situations. We used ornamental grasses and asparagus to create a windswept look."

Multilevel decks break up this yard and give it depth.

Deck continues around pool and links it to the house. It also provides a clean, nonslip surface for poolside fun.

DESIGN FOR A SLOPING SITE

Blum describes the other design. This one uses patios and decks to gain space on a hillside lot:

"The property was made up of very steep grade changes. It was all up and down. The small lawn and patio area was the only flat spot. When the clients came to us they had a dying lawn and a cement pad in that area. They wanted it changed into a pleasant garden where they could spend time outdoors.

"We began by retaining the hill with large pressure-treated fence posts. We regraded just enough to leave room for a small path that moves up the hill.

"The owners wanted a used-brick patio. This blended nicely with their brick house. In our climate we usually try to stay with paving brick for patios. Regular or used brick is too porous. It tends to absorb water, freeze, then fracture. These owners lose about 10% of their bricks that way each year.

"The owners wanted a rustic atmosphere in their yard. To achieve this, we broke up the linear effect created by the narrow corridor between house and hill. The brick herringbone pattern and diagonal placement of deck boards help break up the strong lines of house and hill.

"The overhead is for hanging baskets. It also visually spans the distance between house and hill. It helps make the area more like an outdoor room."

Circular lawn and diagonal brick patio provide visual relief in narrow, linear yard between hillside and house. Retaining walls stabilize the slope for plants and path.

Small deck platform steps from the patio into the house. Arbor provides a place to put hanging baskets.

MASTER PLAN FOR ALL AGES

Jim Gibbs is the owner of Green Brothers Landscape Co. Inc., Smyrna, Georgia. He tells how he helped a family create an outdoor living space to fit their lifestyle:

"The owners had added a new family room to their house. After they added it, they realized they were unhappy with the garden.

"The garden is viewed from almost every window in the back of the house. Because much of it was seen from inside the house, it had to be very pretty. We wanted the garden to invite you outdoors.

"The first problem we faced was drainage. The husband loves to garden but had terrible luck growing anything close to the house. Everything he planted died. I explained to him that we needed to build small berms to elevate the area. We also had to put in an elaborate system of perforated drainpipes to carry away water. The berms also served to break up the large yard into several spaces.

"We had good luck moving most of the plants the owners had invested in. The tall pines were left in place. The husband is a perfectionist and an excellent gardener. He did a great deal of research to identify plants that grow well in shade. His work paid off. The garden seems to have something in bloom almost every month of the year.

"The only other site problem we had was strong sun against the back of the house. It was unbearably hot in the summer. To solve that, we brought in large river birches. They shade the back of the house in summer but let warm sun through in winter when the leaves have fallen.

"I suggested brick as patio material. The circle patterns we used work very well with the berms.

"We knew a lot about the family's lifestyle. It was an important influence on our design. For example, the wife loves to have dinner parties. She usually invites about 12 people. She cooks inside and serves outside. That meant the patio would have to accommodate three or four tables.

"The owners are also very family oriented. They bring over their grandchildren at least once a week. We built a play area out of natural wood. It blends into the rest of the yard nicely.

Children's play area includes monkey bars and wooden support for a tire swing.

Brick patio provides a beautiful garden view from inside the living room. It is spacious enough to seat 12 diners comfortably.

There's also a wonderful indoor-outdoor summer house. This structure is used to store the children's playthings.

"Traffic circulation was very important in the yard. The house had large doors, which opened onto the patio. We wanted to make sure there was easy flow to the rest of the garden once you came out. All the brick walks are 5 feet wide so two people can walk on them together. There is a lawn next to the patio. This provides an overflow area when the owners are giving a large party.

"This project took several years to complete. The owner pitched in whenever he could, but a lot of it was contracted out. We believe strongly in having a master plan for a site. It can be carried out over as long a period as the owner wants. If there isn't a master plan, design features rarely blend together nicely."

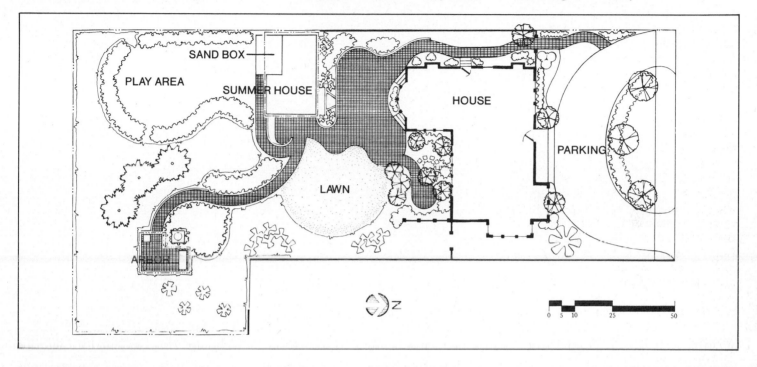

Winding path leads to lower, shady retreat. Path width accommodates two people walking side-by-side.

Curving paths and circular brick patterns complement raised berms. Large deciduous trees and river birches shade windows.

Patio And Deck Ideas

3

By now you should have a good idea of what your outdoor-living requirements are. You should also have a basic plan for the site you'll develop. If so, it's time to look for additional ideas to incorporate into your patio or deck design. On the next 54 pages, you'll find photographs illustrating many different kinds of patios and decks. Use them as a source of inspiration. Borrow a single feature for your design or an entire, attractive theme.

Masters of natural landscaping, the Japanese use *engawas*, small decks, to wed house to garden.

Left: Combination of wood deck, brick patio and stone patio draws people from the house into the yard.

This deck creates a shady spot used solely to watch ships go by. Several trees frame the view.

Patios And Decks Are For People

An important part of designing a patio or deck is planning for activities that will take place there. Consider the leisure pastimes of those who will use the area. Large patios or decks can be divided into several areas. Try to create an environment amenable to various family activities. Sun and shade, privacy and design amenities, such as benches, planters and lighting should be considered.

Notice how the patios and decks on these pages have been planned for the users and their interests. If you have similar requirements, these ideas may be adapted to your patio or deck design.

Above: Large living-room windows and an upper-level deck are designed to make the most of a view, either indoors or out. Deck is covered with indoor-outdoor carpeting, a soft surface for young children.

Left: Stairs lead from the upper deck to a ground-level flagstone patio. Flagstone creates a natural-looking surface which blends well with large trees. Laid on sand, it allows air and water to reach tree roots.

Below: Firepit adjacent to the patio is used for outdoor cooking and spending evenings around the backyard campfire. Stairs that lead to upper-level deck are wide at the base to make the steep climb less intimidating.

Above: Incorporating natural features such as large rocks into the design helps patios and decks fit gracefully into their surroundings.

Left: Large, circular brick patio provides enough room for outdoor entertaining and blends nicely with wooded surroundings. Staggered steps lead to an upper-level pool.

Below: Multilevel decks and brick patio separate yard into different activity areas. Decks also create usuable space on sloping waterfront property.

Above: Wise use of materials and a slightly raised deck define separate activity areas. Diagonal deck boards direct the eye toward the garden.

Left: Multilevel deck provides access to all areas of house and garden, maximizing usable space. Generous use of trees and shrubs softens straight lines created by deck and creates link to wooded surroundings.

Below: Sliding doors, an open arbor and a small brick patio create an intimate connection between indoor and outdoor spaces.

Above: White railing complements house and makes enjoying a spectacular view safe.

Right: Decking in traditional boardwalk style directs traffic and connects different activity areas.

Below: This deck is attached to a docked houseboat. It's a perfect example of how decks can create outdoor-living space where none exists. Flower-filled wooden containers add a bright splash of color.

Create An Indoor-Outdoor Link

Think about how your patio or deck will relate to the house. Indoor-outdoor areas are a practical way to expand living space. An enclosed patio or deck can provide a year-round area for many uses.

Many patio surfacing materials can be used both indoors and out. Using the same material for the patio and the floor of an adjacent room makes a pleasing indoor-outdoor connection. This visual tie-in gives a more open, spacious feeling to a room and helps bring the outdoors inside the house. An important element in the indoor-outdoor link is the use of large windows and sliding-glass doors. The following photos show effective ways to make an indoor-outdoor connection.

Above: House and garden are combined for outdoor living in classic Greek style. Container color stands out against white surroundings.

Left: A glass block overhead, glass walls and marble floors blur the distinction between indoors and out. Geometric blocks and supports complement formal garden.

Below: Sliding-glass doors open to a classic formal patio. Living-room skylight provides light for indoor plants.

This deck is a room without walls. It is modeled after a Japanese farmhouse kitchen. Smoke from the fire drifts up through a hole in the woven-reed roof. A skylight brightens the interior.

Make tea, popcorn, or just keep warm next to the firepit. Pots are suspended by a section of hooked bamboo.

Following Japanese tradition, users remove shoes before going inside. Sliding *shoji* screens open into the house.

This house takes full advantage of its southern exposure. Clear fiberglass roofing traps sunshine and warms the house.

Below: Patio umbrella provides comfortable shade in hot weather. Fireplace can be enjoyed indoors or out.

Above: Dark furniture and carpeting retain heat while reducing glare. Venetian blinds block bright sun if necessary.

Right: Sliding doors and clear plexiglass roofing allow this indoor-outdoor room to be enjoyed any season, rain or shine.

Below: This patio of interlocking pavers illustrates an attractive alternative to rectangular brick. A small, evergreen hedge defines space, separating patio from lawn, but is not confining.

Disappearing pocket doors and continuous Spanish tiles completely eliminate distinctions between indoors and out. The patio is part of the home, and the home is part of the garden.

Below: Large sliding doors open for complete freedom of movement between garden and house. Owners enjoy beautiful, unobstructed view while dining, entertaining or swimming.

Above Left: Patio is comfortably shaded while swimmers enjoy full sun.

Center: Patio tile is a clean, formal surface which works well indoors and out. This tile patio is an attractive complement to the used-brick counter in background.

Below Left: Pocket doors slide on ceiling and floor tracks making a tight seal.

Below: Pocket doors slide neatly out of view into a wall. They are all-weather performers. During good weather, doors are left completely open. Screen doors are pulled out to block flying insects. Glass doors roll out to protect against wind, rain or cold.

Pocket doors slide on ceiling and floor tracks making a tight seal.

Above Left: The combination of plexiglass roof and screen walls allows for outdoor enjoyment free of insects.

Above: Initially, plexiglass panels slid into dadoed 4x4" beams. Expanding and shrinking of the plexiglass resulted in leaks. Dado cuts were replaced by extruded aluminum channels.

Left: Natural look of redwood blends well with the yard's woodsy feel. A hot tub is in the rear. If this family had dogs or small children, a kickboard around the screen's base would be necessary to prevent it from being torn.

Below Left: The hot tub is enclosed in a benched platform for easy access. Platform also hides plumbing and provides storage with outside access. The gentleman at left is the towel attendant.

Precision-cut flagstones make a suitable indoor-outdoor walking surface.

Upper-level deck creates perfect setting for a covered-patio room beneath. Stairs on the side give easy access to either area.

Left: Spacious entryway doubles as an all-weather plant room. On clear days, the roof can be opened hydraulically. Sunlight streams in and hot air out.

Above: Large glass windows open to an enclosed courtyard, giving small rooms a sense of space. The narrow deck/walkway provides new mobility between various parts of the house.

Right: A dramatic use of color, shapes and textures give this enclosed patio a strong Asian flavor. Movable *shoji* screens are space dividers.

Container gardening and patio living are a natural combination. Owners of this home love spring bulbs. The small arbor shades plants and people. Water runs into the drain in the middle of the patio.

This sunny patio is enclosed by a U-shaped home. Large windows bring sight of fresh flowers to every room. Doors provide easy access to patio from anywhere in the house.

A corner overhang was transformed into this protected patio for outdoor dining.

From the inside, diners enjoy the relaxing beauty of natural surroundings.

Courtyards: Patios Privately Linked To The Past

Courtyards are completely enclosed areas adjacent to a house or other building. In most cases, courtyards are a part of the original house design, although they can be added to any house by enclosing an adjacent patio.

Some locations are better suited to courtyards than others. An unused area between two buildings, such as a house and detached garage, can be transformed into a secluded retreat by enclosing the area with walls and providing access from the house and yard.

Courtyards are a part of traditional architecture. The ones shown here offer ideas for both traditional and modern architectural schemes. The feel of a courtyard is not dictated solely by the architectural style of adjacent buildings. You can create almost any architectural theme by careful selection and use of materials and design elements.

Left: A wide variety of foliage textures set off by red brick and tile soften courtyard corners, creating a special place to relax.

Below: The unique, textured beauty of old-brick walls is tastefully highlighted by the freshness of tile.

Courtyards are rich in tradition but difficult to drain. Here, a small canal prevents a permanent puddle in a historic New Orleans courtyard.

Enclosed by four walls, courtyards are naturally secluded and secure. Outside, the world passes by. Inside, an interesting brick pattern draws attention to a center lamppost.

Strong archways are important architectural features of historic New Orleans courtyards. They reflect early Spanish influence.

Right: Courtyards like this exemplify the wonderful intimacy of an enclosed garden area.

Below: This 19x50' courtyard serves several functions. It's an outdoor room for dining and cooking, a stage for an experienced shade gardener and an entryway to the home.

Above: Outdoor-living areas are as individual as the people who use them. This courtyard was designed to be a secluded retreat from a fast-paced world.

Left: Enclosure on four sides creates a special microclimate. This one proved ideal for a collector of rhododendrons and azaleas.

Trees: A Natural Complement

Trees are perhaps the most useful plantings to have around a patio or deck. Deciduous trees provide shade in summer and drop leaves in winter to admit sunlight. In some instances, trees are the only barriers tall enough to provide privacy or a suitable windbreak around the patio or deck.

The photos on these pages show how patios and decks have been planned around existing trees and how new trees have been incorporated in the overall landscape design.

Above: Grading for a level patio often calls for dropping or raising the soil line around roots of an existing tree. This can kill the tree, which then becomes a hazard until felled. Consult a tree specialist for expert advice.

Left: A gnarled old tree becomes a friendly member of the family willing to lend a strong limb for a rope swing. Brick-on-sand paving allows water and air to reach tree roots.

Below: Trees don't have to be old to have effect. Consider making room for new trees in your design. This planter provides enough open soil for free growth and has room left over for annual flowers.

Right: Trees can be functional focal points of a design. Cooling shade and a comfortable place to sit were the result of designing a special well in this raised patio.

Below: Loose aggregate at tree base allows necessary air and water movement into the soil. Designs that place trees out of traffic routes favor tree growth by preventing soil compaction around roots.

Overheads For Enclosure, Shade And Hanging Plants

Patio and deck overheads range from a few decorative open beams to solid roofs. They can be free-standing or attached to the house. As you look at these overheads, study their form and function. Solid overheads provide full shade. Spaced lath or boards provide partial shade. They can act as supports for climbing vines. Overheads of glass, acrylic sheets or translucent-fiberglass panels admit full or filtered sunlight. They protect the deck or patio from rain, snow, and wind. Overheads of widely spaced boards or beams are primarily decorative elements, although they create a feeling of enclosure and support hanging plants.

Above: The angled top of this arbor creates the feeling of an open-air pavilion. It also mirrors similar angles in the house roof.

Left: Separate but part of a plan, a small arbor creates a shade area independent of the patio. Black stain, random black brick and black furniture unify the entire outdoor area.

Below: A slightly raised deck helps identify a separate retreat under the angled portion of the arbor. Brick direction leads the eye to this area and breaks up the straight lines of the arbor.

Above: A deciduous-tree canopy is one of the best covers for a patio or deck. These sycamores provide cool shade in summer, then drop their leaves in fall and allow sun to warm patio and house.

Right: Arbors are easily covered with flowering vines. This wisteria fills the air with fragrance in spring, provides shade in summer and conveniently drops leaves in winter, maximizing sun.

Below: Similar colors and contrasting straight lines link house and arbor architecturally. Arbor blocks sun from deck and living room.

A canvas cover shades just enough of a brick patio to provide a comfortable spot for an afternoon break. Plants and children are happy in full sun.

White paint and pink flowers add special coolness to a shady retreat. Arbor provides a place to hang plants and permits a high, branching tree to spread its limbs.

A fiberglass-covered arbor creates the perfect climate for a collection of tuberous begonias and other shade plants. Bench at right is for repotting and storage.

Right: Large, beamed, multilevel arbor gives a feeling of strength and enclosure to an exposed-aggregate patio adjacent to a small home. Fiberglass top softens the light that enters the area.

Below: Lattice overhead allows fresh air and sunlight to enter enclosed planting area and adjacent living room.

Private Places: For You Only

Small patios and decks can be places for privacy and solitude. A favorite spot in the yard can be transformed into a secluded hideaway by isolating it from the surrounding yard.

Plants, fences or masonry walls can help create privacy. Side yards are good locations for small, private patios. An enclosed patio off the bedroom is another option. Private nooks can be designed into larger patios and decks by using partitions, level changes or extending the patio or deck around a corner of the house.

Right: Careful placement of trees and shrubs makes this off-bedroom deck a private haven within the yard.

Below Right: A one-man deck is an oasis of privacy. Viewed from inside the house, it is a living picture of green foliage.

Below Left: A person's leisure preference determines site suitability. This hot, south-facing corner is miserable for noontime dining but perfect for a sunbather.

Terraces: Bringing The Patio Up To House Level

Terracing is a method of gaining level space on a sloping lot. Terraces on steep hillsides often require retaining walls and safety railings. Multilevel terraces are usually connected by flights of masonry steps or wooden stairways. The natural slope is often left between levels for hillside planting areas. For a more orderly landscape, you can design a series of level planter beds stepping up to the terrace.

Terraces generally require extensive grading. It's often necessary to hire an excavation contractor to do this work. In areas where soil is unstable, enlist the help of a soil engineer or qualified landscape architect to design the terrace.

Below Left: Pieces of broken concrete combine to form a relaxing outdoor area, half patio and half lawn. For circulation, a short path winds down the side of the terrace through a rock garden of dwarf conifers.

Above: A diagonal-herringbone brick terrace rises above the home below, creating a pleasant place to enjoy a beautiful view. Trees offer shade. Cast-cement railings provide security.

Below Right: Small retaining wall converts troublesome sloping land to a level outdoor living area. The striking combination of white and green demonstrates the beauty of simplicity.

Entries Say Welcome!

Entries deserve special attention. They're the first visual impression vistors get of your home. An entry can offer clues to the type of people who live in the house.

Entries also provide a gradual transition between the street and front door. Spacious, attractive entries are desirable because they allow a visitor to pause and relax before knocking on the door.

Above: Decks don't have to be large to be useful. This small display deck accents a narrow side yard and is visible from inside the house.

Left: A modular entryway built of railroad ties and brick doubles as a series of raised planters. It artfully scales a gentle slope. Outdoor lighting provides a nighttime accent.

Below: A small atrium forms a private, secure entryway and patio. Matching color of door and outdoor furniture creates unity.

Above Left: A series of small decks stained to match the house is separated from the driveway by a granite brick path and edging. Piers, rope and lighting lend nautical flavor.

Above: The generous use of seeded-aggregate paving makes a spacious welcome mat and highlights a clump of birch trees.

Left: Pressure-treated timbers and brick climb a gentle slope to the front door.

Above: Straight lines and natural tones of this large entrance deck complement home in color and geometry.

Left: The smallest entryway can be enlivened with container plants. Brick on a sand surface allows for water drainage. Deciduous vine provides summer shade and winter light.

Above: Used brick in basket-weave pattern complements native rock and creates a rustic entryway.

Left: Changing directions and materials, an entryway eases the slope a visitor must climb to reach the front door.

Below: Spacious brick entry is separated from asphalt driveway by granite brick edging. Size and shape of lower landing mirrors the porch.

Side Yards: Making The Most Of A Narrow Area

A side yard can be more than just a place to put the trash cans or store garden tools. It can provide a private spot for an intimate patio or deck, or it can be used for an extension of a larger, backyard patio or deck.

Side yards are usually long, narrow spaces. The photos on this page show how patios and decks can be designed to fit into them.

Right: A narrow side yard adjacent to a mobile home was transformed into a low-maintenance patio with just enough room to grow flowers.

Below: All lines of a side-yard deck and arbor lead guests to a shaded back yard.

Below Right: A canvas-covered side-yard patio provides the perfect climate for shade-loving plants and a relaxing Saturday afternoon.

Right: A combination of several materials forms an inviting entryway. Dark-brown Moroccan brick and white stone highlight the birch trees framing this entry.

Below: Zig-zag brick patio breaks up rectangular side yard and gives it a woodsy personality all its own. Railroad ties are used for raised planters and as hillside steps.

Roof Decks: Outdoor Living With A View

A roof deck is a perfect way to provide outdoor-living space, especially in cramped urban areas. If you have a suitable roof, use these roof-deck designs for ideas.

The first step in designing a roof deck is to find out how much weight the roof will bear. Most of the decks shown here use lightweight materials to decrease the load placed on the roof. Seek advice from a qualified builder or architect before you start.

Right: Lightweight, movable decks allow for a green oasis among the stark walls of the city. Low planters bordering the decking are lined with plastic. They contain a bottom layer of pine boughs for drainage topped by 2 inches of lightweight soil mix. Low-growing or wind-tolerant beach plants are most successful. Drip irrigation provides water.

Below Right: Everything on this roof deck is lightweight, including the feather-light volcanic rock and the woven willow-branch furniture. Plant containers are spread out to avoid concentration of weight in any one area. The Hudson River provides the backdrop.

Below: A vine-covered trellis creates a shady green cave at the end of a deck path. Water exits through drains in the corners of the roof.

Above: A fruit lover uses every inch of his 22nd-floor rooftop deck. Espalier trees save space and add privacy.

Left: Even with enough fruit trees for a small orchard, there is still room for outdoor dining.

Below: This view deck rests directly on an existing gravel roof. 4x4s laid perpendicular to roof joists support 2x2" decking. Benches double as a railing.

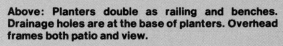

Above: Planters double as railing and benches. Drainage holes are at the base of planters. Overhead frames both patio and view.

Left: Diners enjoy a spectacular view when eating on an easy-to-clean wooden table. A small planter box acts as a centerpiece.

Left: Wooden fence offers privacy on this low rooftop deck. A tall tree planted at ground level provides shade.

Below Left: This is a completely modular roof deck. The wooden structures are easily dismantled and moved.

Below Right: A tiled rooftop is transformed into a green island in New York City. Container plants are grouped to create separate areas.

Benches Are More Than Just A Place To Sit

Benches can be decorative and functional elements of a patio or deck. They can double as railings or display areas for plants or other objects. Benches can also serve as space dividers to define activity areas within the patio or deck. For more information on designing benches, see page 155.

Above: A hinged-top bench doubles as weather-tight storage unit for boating equipment.

Left: Thoughtful design leaves no hint that this bench top covers a storage compartment.

Below: Swinging benches are one of the most relaxing forms of seating. They require strong support. This one uses pressure-treated posts embedded in a cement pad.

Above: Good bench design allows for easy, face-to-face conversation. Bench supports are set in concrete and raised off the patio surface for a sturdy bench and easy cleaning.

Right: A bench can double as a protective railing for a roof deck. This one also serves as a display area for the owner's bonsai collection.

Below: Benches are important design elements. Here, they help break up the expansiveness of a large deck, define activity areas and act as railings.

Above: This bench is supported by large blocks toenailed to the deck. It also provides a safe railing for toddlers.

Above Right: Benches can also be the starting point for an exciting swing.

Left: Prefabricated-metal supports make a strong, easy-to-install bench. A 4x4" fascia around the bench top lends cleaner, more styled look.

Below Left: Benches must be strong enough to support the weight of several people. The supports for this bench are well-designed for that purpose.

Below: Benches can be used to display anything from plants to modern art.

Railings For Safety, Separation And Style

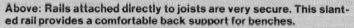

Wooden railings give decks a finished appearance. Railings are a necessary safety feature on decks or terraced patios more than 2 or 3 feet above ground level. Railings can serve as backrests for benches or as act as benches themselves. For more information on designing railings, see page 153.

Above: Rails attached directly to joists are very secure. This slanted rail provides a comfortable back support for benches.

Below Left: Rail posts on this deck are capped with ornaments, called *finials*. Finials can be wood, metal or ceramic.

Below: Extended structural posts are strong railing supports. Here, rails are set into slots, called *dadoes*, cut into the posts.

Above: This bench is made of 2x4s laid on edge. Its design complements the railing on this hillside deck.

Right: Rail designs are limitless. A rail made of ebony-stained posts provides a sharp contrast to the glossy, natural finish of an fine-crafted deck.

Below: Close-set balusters on railing provide safety on this high-level deck.

Left: Contrasting colors highlight a railing that also screens the foundation of a raised deck.

Below: Rail posts must be securely attached to the deck's main frame to ensure a safe place to lean.

Steps And Ramps To Link Levels

Steps should be easy to climb. They should also complement the patio or deck design. Steps can be made of wood, masonry materials or a combination of the two. Outdoor steps are usually designed on a larger scale than indoor stairs or steps.

Ramps are a practical alternative to steps if you plan to move wheeled vehicles and devices on and off the patio or deck. For more information on designing steps and ramps, see pages 27 and 155.

Above: This wide flight of steps was made by pouring concrete into redwood forms. The surface was then finished to expose the aggregate in the concrete. For details on exposed-aggregate surfaces, see page 113.

Left: Steps provide easy access from an elevated home to a lower-level brick patio. Strong railing offers safety and support.

Below: Steps made from 4x4s ascend a steep rise to connect two activity areas.

Above: Black wrought-iron railings provide safe, visible support to a steep brick stairway.

Left: Wooden ramps offer a smooth alternative to steps. They provide access for wheeled vehicles.

Right: A nautical ramp connects this houseboat to the dock. Rollers and metal plates allow the houseboat to move with the tides without harming the dock.

Below: This circular staircase rises to a great height without occupying much space, a distinct advantage in small areas. Wrought-iron rail matches the patio furniture.

Above: Three sections of stairs provide access to three areas of home and yard. Stone walls and naturally weathered wood make a pleasing combination.

Left: Small spaces between these steps allow just enough room for a creeping vine. The vine softens the sharp lines of the brick. The steps are wide enough to accommodate several people at once.

Above: These beam-and-brick steps allow two people to walk side-by-side, or pass going opposite directions. Proper spacing between steps makes for an easy climb.

Right: This series of small deck platforms supported by round piers forms a layered set of stairs with a nautical feel.

Below: Stairs connect every level of this home and terraced garden, providing easy access to all areas.

Paths Invite, Connect And Direct

When planning paths into the landscape, consider convenience as well as beauty. A curving path adds more interest to the yard than a straight one, but a path that winds too much will be inconvenient to those in a hurry.

A number of masonry materials are suitable for paths. Steppingstones or wooden rounds can be used for directing people through low plantings or across lawns. The materials you use should make a smooth walking surface. Loose aggregates, such as gravel or wood chips, will need occasional replenishing.

Above: Sweeping curves of stone and wood lead to a secluded arbor. Generous path width prevents traffic congestion.

Right: Mounded soil, dense greenery and a flowing brick pattern create a path which beckons toward a larger patio. This spacious patio is designed for outdoor entertaining and dining. There is room for three sets of tables and chairs and movement in between them.

Below: Several activity areas of this gently sloping yard are linked with exposed-aggregate steps. Lights allow for safe movement at night. Arbor/leisure area is also a garden storage center.

Left: Round, white steppingstones provide a direct link to this circular white guest house. This is a beautiful example of reflecting house architecture in the paving design.

Below Left: A combination of stone and old wood result in a natural-looking path. It is designed to simulate a dry creek bed. A single outdoor light stands in contrast to the rustic design.

Below: Sunken beams soften a large expanse of asphalt, link it to its wooden surroundings and provide a path from home to yard.

Above Left: Wood rounds embedded in the lawn link this tile patio to the rest of the garden. When sunk to the proper depth, rounds do not obstruct lawnmowers.

Above: Granite block pavers provide easy access from front yard to back.

Left: Brick path, straight-edged hedge and tall walls boldly point to the only exit in this enclosed courtyard.

Below: Brick edged with pressure-treated beams form sharp, clean lines on this walkway.

Here are three paths with three different textures—seeded river rock, left, brick, above, and steppingstones in loose aggregate, below left.

Wood rounds, old wood beams and rocks of every shape and size form this path connecting outdoor aviaries.

Wooden Decks For Pools And Tubs

Above: An old, rarely used above-ground pool was given new life when completely enclosed with a weathered redwood deck and poolside seating.

Left: A birdseye view of the same pool shows how the deck area was integrated into a narrow waterfront yard.

Below Left: Similar sharp angles in house, pool, exposed-aggregate patio and deck create oneness. The effect is a result of thoughtful design.

Below Right: An irregular-shaped pool surrounded by wood and stone has the feel of a country swimmin' hole.

A wooden deck can provide pleasing contrast to a masonry swimming pool or spa. Poolside decks help break up large expanses of concrete. Wooden decks are an obvious choice around hot tubs.

Use decay-resistant wood for decks near the water. The best woods are those that have been pressure-treated with a wood preservative. In addition, the deck surface should be treated with a water-repellent sealer or finish. See pages 136 and 157.

Above: Multilevel deck and steps were designed to provide easy access to this hot tub.

Left: Raised deck lessens visual impact of a white diving board in a natural rock pool.

Patio Builder's Guide

Patios have important historical significance to both architectural and social aspects of modern outdoor living. Their origin can be traced to early Spanish courtyards that served as centers for family activity. They have evolved through generations until the word "patio" means almost any outdoor activity area, including wooden decks.

This chapter groups patios into outdoor surfaces installed at ground level. With the exception of wood, all patios discussed here are masonry. Each masonry material requires specialized construction techniques. Installation details accompany each discussion.

Patios can be installed on stable soil, a bed of sand, sand and gravel or a poured-concrete pad. Patios differ from decks in one important way—because patios are at ground level, you must pay particular attention to water drainage. While a deck can be built above ground to avoid drainage problems, a poorly planned patio more often creates them.

Left: Cut slate makes a formal-looking flagstone patio. This slate patio is being laid on a sand bed. The worker is mortaring patio edges to keep pieces from shifting. Slate provides a durable, if slightly rough, patio surface. For information on installing flagstone patios, see page 117.

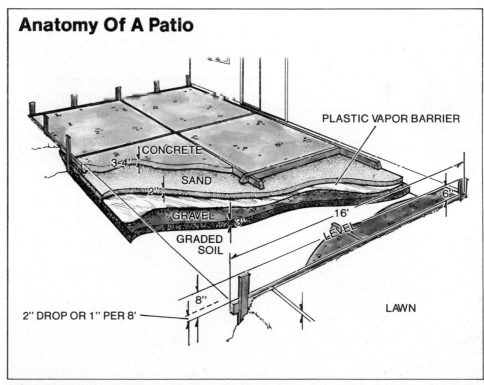

Anatomy Of A Patio

PLASTIC VAPOR BARRIER

CONCRETE

3-4"

SAND

2"

GRAVEL 3"

6"

16'

GRADED SOIL

LEVEL

8"

2" DROP OR 1" PER 8'

LAWN

Cross section of completed concrete patio shows basic foundation elements. Your patio may or may not require all of these elements, depending on the surface material you use and the soil stability of the site. This drawing also shows the recommended slope for a concrete patio—1 inch drop per 6 to 8 feet of length. Slope patio to drain away from house.

Determining the slope, or *grade* and planning for water movement is the single most important step in patio construction. Planning for grading and drainage is discussed in the chapter "Planning Your Patio Or Deck," pages 24-26. Basic grading and drainage considerations for patios are discussed in the following pages. Plan carefully and read through the entire construction sequence before you begin.

Excavating The Patio Site

The first step in patio construction is *excavating* the site. Excavation involves removing loose topsoil so the patio will have a firm base. Excavation is also necessary to install the finished patio at the proper level. During the excavation process, you'll be *grading,* or sloping, the ground for water drainage. Grading the patio site for proper drainage is one of the most important steps in construction. Excavation steps are as follows:

1. Use string and batter boards to outline the patio area. Level the string with a line level. Make sure the patio area is square by using the "3-4-5" triangulation method, described on page 140.

If you're tearing out an existing lawn, remove sod 3 inches past the patio outline strings. If you want to save the sod, cut it into strips 16 inches wide by 5 feet long. Roll up the strips and keep them damp until you replant them.

2. Drive 2x4 grade stakes at 3-to 6-foot intervals around the patio perimeter, as shown in the drawing. If you're pouring concrete, position stakes so they can also serve as supports for wooden forms.

3. Determine the excavation depth by adding thicknesses of patio surface material and foundation materials. The thicknesses of foundation materials will depend on the stability of the subsoil, as described in "Site Drainage And Grading" on the facing page.

4. Decide which way you want water to drain off the patio surface. If the patio is next to the house, water should drain away from the house. Refer to page 26 for determining the correct grade for different patio surfaces. For smooth concrete patios, allow a 1-inch drop for every 6 to 8 feet of patio.

5. To determine grade, start at a corner grade stake on the proposed high end of the patio. On the stake, mark the proposed height of the finished patio surface.

Measure down from the leveled outline string to the mark on the grade stake and note the measurement. Mark this distance, plus the drop in grade, on the grade stake at the low end of the patio, measuring down from the leveled outline string.

6. Attach a chalk line to the inside faces of end grade stakes at the marks. Snap the line across intermediate stakes. Align the top of the 2x4 or 2x6 form to the marks on the stakes. Attach the form to the stakes. Repeat this entire procedure at the opposite side of the patio area. If you have a large patio area, you may have to add grade stakes across the center to ensure proper grade throughout.

7. Install forms for the remaining two sides of the patio area and start excavating. Use the tops of forms as guides to dig to the correct depth. The finished patio surface should be even with the form tops. Use a square-point shovel to dig excavation edges flush with forms.

8. If the soil at the bottom of the excavation is loose or spongy, you'll need to compact it. Compact soil by moistening it and tamping it with a tool called a *tamper.* This will prevent uneven settling after the patio is installed. Lay the sand or sand and gravel base, being sure you maintain the correct grade. Tamp and level the foundation surface. You are now ready to install the patio.

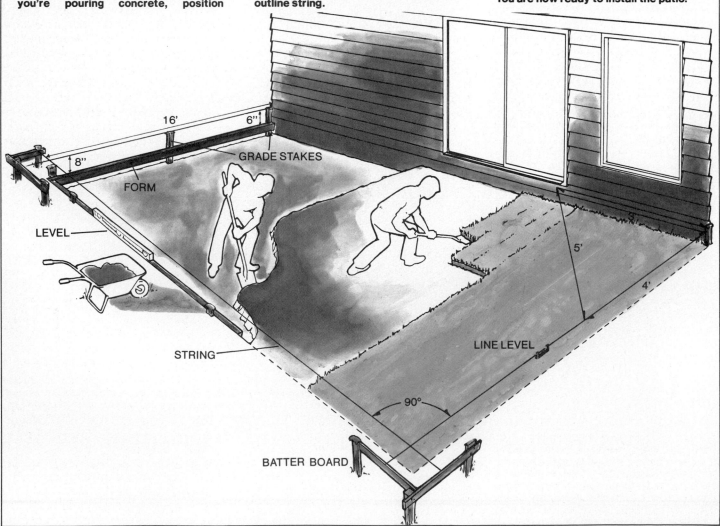

Preparing To Build

Many patio surface materials offer choices of texture and style. The following sections in this chapter describe the different materials. They will help you choose the right one and install it properly.

SITE DRAINAGE AND GRADING

Grade the patio to drain away from the house. The slope will be determined by the patio material you use. See page 26. After laying out the shape of the patio, determine the elevation. Use strings tied to stakes or batter boards as guides while grading. It's a good idea to make the patio surface flush with the ground to prevent people from tripping over raised edges. It's easier to mow and trim a lawn around flush patios.

The ground stability after topsoil is removed will determine the type of base required for the patio. On stable, well-drained soil, a 2-inch base of sand is adequate for most surfaces. Some materials, such as flagstones or adobe bricks, can be laid directly on stable soil. If you have soggy soil, or hard clay soil with poor drainage, lay a base of 3-4 inches of pea gravel topped by 2 inches of tamped sand. Use this base where soil is subject to shifting from freezing and thawing.

If the site has poor drainage, lay a drain line to carry off water. See pages 25-26 for details. Tree or plant roots can grow and crack your patio. Remove potentially harmful ones while grading. If you want to save a nearby tree and are uncertain of the extent of its potential root growth, consult a nurseryman or tree surgeon.

If you're placing water or electrical lines under the patio, lay them before installing the base materials. Follow code requirements for correct depth of underground lines.

If termites are a potential problem, apply insecticide to the foundation just before you lay the finish patio materials. This helps protect adjoining wood surfaces. Consult an exterminator for advice on which insecticide to use and how to apply it.

TOOLS

You may already have many of the tools needed to build a patio. You'll probably have to rent or buy certain specialized tools. Tools for building patios are shown below.

You'll need a shovel for excavating. Round-point shovels work better for breaking up hard soil. A square-point shovel is better for dirt removal and making a hole with straight sides. A pick is useful in extremely hard or rocky soil. Use a tamper for compacting soil underneath the patio.

For wooden edgings, forms and batter boards, you'll need the following carpentry tools: *hammer, nails, saw, mason's twine* or *heavy string,*

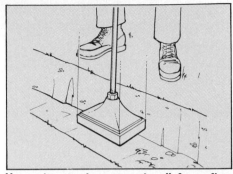

Use a tamper to compact soil for a firm base on which to lay the patio foundation. The metal tamper shown here can be rented at a tool rental outlet. You can make a tamper by nailing a 1-foot piece of 2x8 or several thicknesses of plywood to a 4x4.

plumb bob, carpenter's level, line level and a *framing square.* Other helpful tools include a *folding rule* or *steel tape measure* and a *chalk line.*

Masonry Tools—For bricks, tile, stone and other individual masonry units, you'll need a pointed *mason's trowel* with a 10-inch blade for applying mortar. To cut bricks and rocks, you'll need a *mason's hammer* and a broad-blade *cold chisel* or *brick set* to cut these materials and smooth their edges. A brick set can also be used to cut large patio or quarry tiles and flagstone. For smaller tiles, you can use a *tile cutter,* available at tile suppliers. To shape or *strike* mortar joints, you'll need a *striking tool.* Striking tools come with different shape tips to form the desired mortar joint shape.

For concrete surfacing, you'll need a large *bull float* and a smaller wooden *cement float,* called a *darby,* for initial surfacing and smoothing of concrete. For smooth-finishing concrete, you'll need a steel-blade, rectangular *fanning trowel.* This trowel gives the surface a smooth finish. Initial leveling of poured concrete is done with a *screed,* usually a scrap piece of 2x4. See the photo on page 111. A screed can also be used for leveling sand, gravel, or soil before the patio is installed. You'll need an *edger* for rounding the top edge of the concrete patio, and a rectangular trowel, called a *jointer,* to cut control joints or expansion joints in the concrete surface. A wheelbarrow and hoe work well for mixing small amounts of concrete. For large patios, it's generally more practical to order concrete wet-mixed and have it delivered. A pair of rubber boots and gloves will help you keep clean.

Heavy equipment for excavating and motorized cement mixers can be rented from a tool rental outlet.

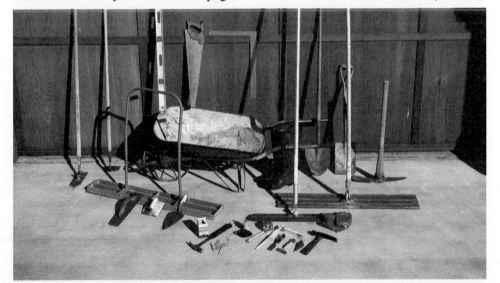

Shown here are some of the tools commonly used to construct patios: Wood float, metal edger for rounding concrete patio edges, pointed mason's trowel, hammer and nails, chalk line and chalk, metal tape measure and folding rule, wood chisel, brick set or mason's chisel for cutting brick or stone, plumb line, mason's hammer, jitterbug for leveling and compacting concrete, bull floats, heavy gloves, jointer for making control joints in concrete, hoe, mason's levels in 4-, 6- and 8-foot lengths, wheelbarrow, handsaw, carpenter's square, rubber boots, round-point shovel, square-point shovel and pick. The tools you'll need will depend on the type of patio you're building.

Used brick makes an attractive, informal patio. Moss-encrusted mortar joints add to the rustic charm of this patio.

Brick

Brick is probably the most popular material for outdoor flooring. It has been used for centuries to pave streets, walkways, courtyards and garden sitting areas. The appeal of its earthy texture and color is universal. Brick is a good patio material for the do-it-yourselfer. Individual bricks are easy to handle and the work can be done in phases, over a period of time.

Advantages
- Uniform size and shape makes design and installation simple.
- Available in many shapes and colors.
- Can be laid in different patterns to create different effects.
- Absorbs water. Cools surface as water evaporates.
- Durable. Can be laid over a variety of subsurfaces.
- Small sections are easily replaced.

Disadvantages
- Absorbed oils can leave permanent stains.
- Cold weather can cause cracking and heaving.
- Rough surface makes some activities such as dancing and riding children's toys difficult.

KINDS OF BRICK

There are thousands of shapes, textures and colors of brick. You should have no trouble finding a brick type and pattern that will suit your taste.

Common Brick—Also called *building brick,* is the type most often used for patio surfaces. It is porous and relatively rough in texture. Brick color, texture and hardness is determined by the color of the clay used and the amount of heat the brick receives during firing. Brick dimensions vary, depending on the manufacturer. Generally, common bricks are 2-1/4x4x8''.

Ordinary red clay is used most often for common brick. Colors range from hard-burned dark red to softer, under-burned salmon pink. *Clinker* bricks are rough looking with dark blotches on their surfaces caused from over-burning. Also called *hard-burned bricks,* they are the hardest of the common brick types and best resist chipping and weathering. Unless you object to their appearance, use clinker bricks for a durable, long-lasting patio surface. Salmon bricks, also called *green bricks,* have received the least amount of firing and therefore are most susceptible to chipping, breaking and weathering. These bricks are not necessarily green or salmon-colored. You can tell them not only by their lighter color but also by the sound they make when tapped. A clinker brick makes a ringing sound when tapped; a green brick does not.

The surface texture of common brick depends on how the brick is formed. *Wire-cut* bricks have sharp corners and rough, pitted faces. The sides are smooth. *Sand-mold* bricks have smoother faces than wire-cut bricks. The sides of these bricks are slightly tapered by the molding process.

Face Brick—This type of brick is tougher and smoother than common brick. It is often used where a finished or formal appearance is desired, such as for fireplaces or building facades. It is more expensive than common brick and not as widely available. Because face brick has a slicker surface than common brick, it is not recommended for walking surfaces in wet areas, such as around a hot tub.

Used Brick—These are face bricks or common bricks salvaged from old buildings or other demolition. Their surfaces may be worn or scarred and are usually covered with old mortar or paint. Used bricks make an attractive, informal patio surface. Some manufacturers simulate used brick by scarring new brick and splattering it with mortar.

Pavers—These come in a variety of shapes, sizes, colors and textures. Pavers are usually half the thickness of common bricks, sometimes thinner.

All brick is rated for durability. SW—severe weather—bricks are suggested for outdoor use. In climates not subjected to extreme weather conditions, MW—moderate weather—bricks may be used. Never use NW—no weather—bricks outdoors; they won't endure outdoor conditions.

BRICK PATTERNS

Basic brick-laying patterns can be mixed or changed to suit personal tastes. Some common patterns are *running bond, jack-on-jack, basket weave, diagonal herringbone* and *whorling square.* Variations on the basket-weave pattern include *half basket* and *double basket.* These patterns give an ordered, consistent appearance that is pleasing to the eye. You should not mix patterns too much in small areas. This can be more confusing than pleasing to the eye. In large areas, varied patterns can break visual monotony. The patterns shown in the photos at right are simple to install. With the exception of the herringbone pattern, brick cutting is minimal.

You can easily estimate the number of standard-size, 4x8'' bricks needed to cover your patio. You will need five bricks per square foot if bricks are laid flat in square or rectangular patterns. For curved patios or brick patterns that require much brick cutting, plan on six to seven bricks per square foot. Bricks are sold by individual units or in lots of 500 or 1,000 units. When estimating brick numbers, figure 5-10% extra for breakage.

LAYING BRICK

Brick's biggest advantage is its small, uniform unit size. This makes it possible to install the patio in sections as time permits. Paving in small sections is especially easy when you are laying bricks on sand. You can stop at any point without worrying about a mortar or concrete base setting up before you finish a section.

Laying Bricks On A Sand Base—After you've excavated and graded the site, use a tamper to tamp the soil firmly. Construct a form of 2x4s around the perimeter of the excavation. The height between the top of the form and the bottom of the excavation should be about 5 inches. For curved patio edges, make the form out of 1/2x4'' benderboard. You can either leave the forms as a permanent edging or remove them after the bricks are laid.

Spread a 2-inch layer of sand over the patio area. Wet the sand to settle it. Construct a screed from a length of 2x4 and a shorter length of 1x4. Cut the 1x4 to fit between the forms, the 2x4 to fit over the tops of the forms as shown in the photo on page 107. Nail the two pieces together so the screed will level the sand to the desired height below the forms. If the forms are to be permanent edgings flush with the patio surface, level the sand about 2-1/2 inches below the forms. See the photo sequence on page 107. The screed should be no more than 6 feet wide. You should level only about 12 to 18 square feet of surface at a time. Level the sand by pulling the screed toward you.

Tamp the sand. If necessary, add more sand to maintain the desired level and repeat the screeding-tamping procedure.

Usually, bricks are set on sand with sides butted snugly together. You can space bricks slightly but the resulting surface will not be as stable, unless you mortar the joints. If your bricks vary in size, open joints may be necessary.

To lay bricks, start at one side of the patio. Lay one course, or row, of the pattern, starting at the center of one side and working toward each corner. When you get close to either end, adjust spacing so the course will end with full bricks, if possible.

Use a rubber mallet or a hammer and wood block to gently tap the bricks into the sand, as shown in the photo on page 107. As you lay bricks, check each row with a level to make sure you're maintaining the proper grade. Use a straight length of 2x4 as a straightedge to make sure all bricks are at the same height.

The bricks will settle, so set them a little high. Don't kneel on the bricks you have just laid.

After the bricks are all laid, use a level to recheck the grade. Then spread a layer of fine sand over the surface. If the sand is moist, let it dry for a few hours. Sweep the sand into the cracks. Repeat this procedure

Brick Patterns

DIAGONAL HERRINGBONE

HERRINGBONE

TRADITIONAL

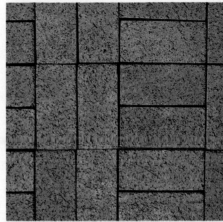
BASKET WEAVE VARIATION

WHORLING SQUARE

JACK-ON-JACK

LADDER WEAVE

HALF BASKET

until cracks are filled. Spray a mist of water on the patio to settle the sand in the joints. For butted joints, you'll need at least 9 cubic feet of sand per 100 square feet of surface. For 1/2-inch joints you'll need 13 cubic feet of sand per 100 square feet.

DRY-MORTAR PROCESS

If you want a patio with open joints, use a dry-mortar process to set bricks and mortar between them. Instead of using plain sand as a base, use a mixture of 1 part Portland cement to 6 parts sand. Then lay the bricks with 1/2-inch joints.

Next, spread a dry mixture of 1 part cement to 4 parts plastering sand across the bricks. Plastering sand is finer than the sand you'll be using for laying the base. It's available in sacks at a sand and gravel yard. Sweep this mixture into the cracks with a hand brush or whisk broom. Tamp the sand-mortar mixture between bricks with a 4x6x1/2" board, being careful not to misalign bricks. Then sweep the entire surface with a broom. Work gently and leave the bricks as clean as possible.

Spray the patio with a fine water mist to remove any mortar still on the brick faces. Be careful not to dislodge mortar in the joints, which will stain the bricks.

As the mortar starts to harden, use a 1/2-inch dowel or a striking tool to finish the joints. This procedure is called *striking* or *raking* the joints. Allow the mortar to set for two hours. Then scrub each brick with a wet burlap bag and let dry. If there are any mortar stains, soak them with water. Then scrub them with a mixture of 1 part muriatic acid to 9 parts water. Use rubber gloves. Rinse the bricks thoroughly to prevent acid stains.

In either process, if you have to stand or kneel on finished rows of bricks, use a large sheet of plywood to distribute your weight over them. This procedure will prevent bricks from shifting.

LAYING BRICKS IN WET MORTAR

You may need a more stable foundation because of shifting soil or severe winter conditions. If so, lay bricks in wet mortar on a concrete slab poured over gravel. See page 108 for information on pouring concrete. Once the concrete slab is poured and hardened, thoroughly wet the slab

and bricks a few hours ahead of time. This keeps them from drawing moisture out of the mortar you will spread on the slab. Mix the mortar according to directions on the bag. Spread a 1/2-inch-thick layer of mortar over a portion of the dampened concrete slab. Pay special attention to the drying time of the mortar. If you spread too much mortar, it may set before you have time to lay all the bricks over it.

Lay the bricks over the mortar, leaving 1/2-inch joints between the bricks. Tap and level bricks as they are laid. Let the mortar set for four hours. Using a pointed trowel, pack the joints with wet mortar mix—4 parts plastering sand, 1 part cement. Let the mortar set for about 1/2 hour. Then finish the mortar joints by striking them off with a 1/2-inch dowel or striking tool. Use a piece of plywood to work on to distribute your weight over the bricks. After the mortar has set, clean the bricks as described in the dry-mortar process, at left.

If you want to lay bricks over an existing concrete patio without raising the patio height too much, use pavers. These bricks are about half the thickness of regular bricks. There is a special epoxy available for bonding bricks to finished concrete. Check with local home-improvement centers or sand and gravel dealers. If you use epoxy, remember that it is hard to clean off exposed surfaces.

If you use mortar, follow the procedures described above. Wet the concrete slab thoroughly before applying mortar. Spread the mortar a little thicker than you would for a new slab. This will create a better bond and allow you to adjust bricks for proper surface drainage. Allow bricks to set overnight before mortaring the joints. Be certain the concrete slab is in good condition before laying bricks.

CUTTING BRICKS

When laying a brick patio, you'll most likely have to cut some of the bricks, especially if you use an overlapping pattern. First, loosely lay all the whole bricks in their correct positions over the entire surface. Then you can see exactly what sizes and shapes you need to cut. Cutting them all at once is most efficient. Once you have the bricks cut, mark them with chalk or a pencil to indicate where they go. Then you can set all bricks permanently.

To cut a brick, use a *brick set,* which

is a broad cold chisel. Its blade usually has a beveled side, which should be positioned away from you. It's best to *score* the brick both sides with the brick set before cutting the brick. Do this by tapping the brick set lightly with a mallet to make a shallow groove or indentation in the brick surface. Then cut the brick, striking it sharply with the brick set and hammer. Rough brick edges can be smoothed with a chisel or mason's hammer.

To cut a brick, use a hammer and brick set to score both sides of the brick where you want it to break, top. Then tap the brick sharply along the score line to break it, bottom.

How To Lay Bricks On Sand

Use a straight 2x4 and a level to check level in both directions across the forms. Allow a slight slope in one direction for drainage.

After firmly tamping the soil, pour the sand base for the bricks. Level sand between the forms with a screed. This screed was constructed from a 1x4 and 2x4 to fit between the forms and level the sand a brick's width below them.

On the leveled sand bed, place bricks in the desired pattern.

Use a hammer and a short block to gently tap the bricks into place. Check surface level frequently with a mason's level.

Pour clean, dry sand over the finished brick surface to fill the cracks.

Sweep sand into the brick joints, then remove excess sand on the patio surface.

Completed brick patio surface is level with the tops of permanent wooden edgings. Note how bricks were cut to fit around existing rocks.

Concrete patio makes a smooth, long-lasting walking surface. Control joints in concrete help prevent cracks across the surface.

Concrete

Concrete is a mixture of water, Portland cement, sand and gravel. Yet it needn't form the bare stretch of flat gray you might expect. Concrete can produce a surprising variety of patio surfaces. Patterns, textures and colors are seemingly endless. You may use it alone or in a multitude of combinations with other materials. It can make a beautiful, economical and extremely durable patio fitted exactly to your tastes and needs.

Advantages
● Variety in surface textures make concrete suitable for a number of outdoor activities.
● Very durable and requires little maintenance.
● Can be formed and colored to simulate other masonry materials.
● Relatively cheap and readily available.
● Can be installed quickly.

Disadvantages
● Takes careful thought and planning—no room for error.
● Difficult to work with—requires some specialized tools.
● Work must be done quickly.
● Large expanses can be monotonous. Should be broken up with other materials or by varying textures of different sections.
● Unless colored or painted over, the surface can be glaring and hot.

More than any other patio surfacing material, concrete requires careful planning ahead. Laying concrete has to be done quickly, precisely and by the rules. Sloppy site preparation or finishing may mean later repair or replacement. Once you begin laying concrete, you must follow through and work fast.

BUYING CONCRETE
There are three ways to buy concrete. How much money and effort you are willing to invest will determine your choice.

The least-expensive but most-demanding way is to buy individual components and mix them yourself. Concrete is a mixture of Portland cement, sand and gravel. The mixture is called *paste*. The dry paste is mixed with water to make wet concrete. Sacks of *dry ready-mix* cost more than components bought separately, but ready-mix can save you time if you're paving a small area. For a very large area you may want to order *wet-mix,* or *transit-mix,* delivered by truck. Some companies rent trailers containing about a cubic yard of wet-mix. You haul it with your own car or truck. Wet-mix is less expensive than dry ready-mix. Wet-mix involves less work on your part, especially if you have a large area to cover.

MIXING CONCRETE
A standard formula for paving concrete is known as *6-gallon paste*. It consists of one 94-pound sack of cement, 2-1/2 cubic feet of damp sand and 3-1/2 cubic feet of 1/2-inch diameter gravel or crushed stone. To this paste mixture, add 5-1/2 gallons of water. If the sand is wet, add only 5 gallons of water. If it is very wet, add only 4-1/2 gallons. To determine wetness of sand, squeeze it in your hand. Damp sand falls apart. Wet sand forms a ball. Very wet sand forms a muddy ball that leaves sand sticking to your hand.

This recipe will make approximately 5 cubic feet of concrete, or about 20 square feet of concrete 3 inches thick. If you're mixing your own concrete, this is close to the maximum amount you can mix in a wheelbarrow.

The standard thickness for concrete patios is 3 inches. Driveways, slab house foundations and other concrete surfaces that must bear heavy loads are 4 to 6 inches thick.

Another way to state the above formula is 1 part cement, 1/2 part water, 2-1/2 parts sand and 2-3/4 parts gravel. Keep this ratio constant for all size batches. All components should be clean and dirt free. Sand and gravel dealers sell clean, washed sand especially for use in concrete mixes. Gravel should be washed thoroughly before mixing. Use clean water for mixing.

If the soil is subject to freezing and thawing, use Type 1-A Portland cement containing an *air-entraining agent*. As the concrete hardens, this chemical forms billions of microscopic bubbles. The bubbles cushion the con-

You can easily mix up to 5 cubic feet of concrete in a wheelbarrow. For larger batches, rent a motor-driven cement mixer or buy wet-mix concrete delivered by truck or trailer.

crete against freeze and thaw stresses. Light-gage wire mesh, such as 8/8 or 10/10 gage should be used for reinforcing patios in cold climates. Reinforcing mesh reduces concrete cracking due to stress caused by heaving as ground freezes and thaws. For steps and driveways 6/6 mesh or 1/2-inch metal reinforcing rods should be used. Air-entrained concrete must be mixed by machine. Hand-mixing isn't vigorous enough to agitate the air-entraining additive.

BUILDING CONCRETE FORMS

Grade and level the site to allow proper drainage. See page 103. Because poured concrete is a solid, one-piece surface, you must provide good drainage underneath the slab or it will lift and crack. If soil drainage is poor, lay a base consisting of 3-4 inches of gravel covered with 2 inches of damp sand. Use a tamper to compact the sand before pouring the concrete.

Concrete is heavy and pushes its weight around as it is poured. You'll need sturdy, well-braced forms to contain it. Forms can be made of 2x4s or 2x6s, depending on the thickness of the slab. Cut stakes from the same material. Drive them in the ground so their tops are slightly below those of the forms. Tack the forms to them with 10d duplex nails. The ''d'' represents nail *penny size* as described on page 135.

If the forms are to act as permanent edgings or joints for the patio, use a decay-resistant wood, such as cedar, redwood or pressure-treated wood. Stake them with 2x4s of the same material. Space stakes no more than 4 feet apart. Drive the stakes about an inch below the tops of the forms so the stakes can be covered by sod or cement. Cut stake tops so they angle away from the boards. Permanently attach forms to the stake with 10d galvanized nails. Cover permanent forms with masking tape so they won't be stained by the concrete.

While building the forms, frequently check them with a level to make sure they follow the desired grade. To build curved forms, use 3/8-inch plywood, benderboards or 1x4s. Benderboards are thin strips of wood used for curved garden edgings. They're available at most nurseries and lumberyards.

If you're using 1x4s, make 3/8 to 1/2-inch saw cuts, called *kerfs,* in one

A concrete patio is durable and easy to keep clean. Here, wooden overhead and house are painted to match bare concrete color. The yard is designed in combinations of square and rectangular surfaces.

side of the board. Space kerfs about 1 inch apart along the board where it will curve.

Drive stakes around the curve outline of the patio so the curved form can be weaved through and attached to them. Set stakes 1 to 2 feet apart on short, tight curves, 2 to 3 feet apart on long curves. Wet the forms before bending them.

CONTROLLING CRACKS

Concrete expands and contracts with temperature fluctuations. To control concrete cracking, provide control joints, also called *expansion joints* or *crack lines.* Space control joints 10 feet apart in patios, 5 feet apart in walkways. Control joints don't prevent cracking. Cracks develop along the joints rather than across the patio surface. To make control joints, use the *jointer* described on page 103. Lay a 2x4 across the tops of the forms over the joint line. Use it to guide the jointer.

Concrete is less likely to crack if it is poured in small sections. You can divide large surfaces into smaller sections by using permanent wooden dividers the same size as the wooden forms or edgings. The forms should be of a decay-resistant wood, such as redwood, or pressure-treated wood.

If you will be working alone, use either permanent or temporary wooden forms to divide the patio into workable sections. Once you pour concrete, you must level, float, edge and finish-surface it before it sets. Mix only as much concrete as you can use. A good amount is 5 to 6 cubic feet. This amount will fill an average-size wheelbarrow or motorized cement mixer. The formula on the facing page will make 5 cubic feet of concrete. Start with smaller batches than this until you're able to gage your working speed.

If you've ordered a large quantity of wet-mix, it's best to have several helpers. Build forms before the wet-mix

Typical Concrete Patio Cross Section

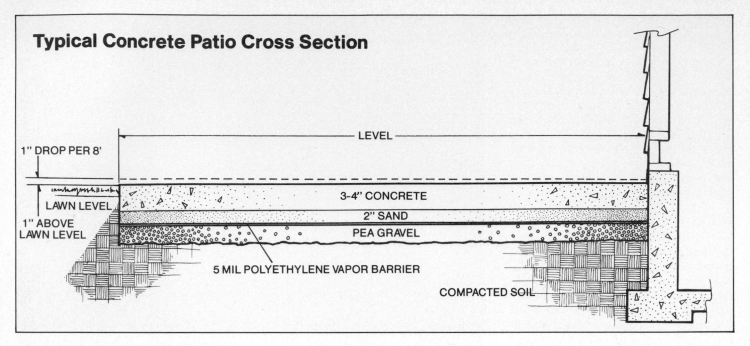

LEVEL

1" DROP PER 8'

LAWN LEVEL

1" ABOVE
LAWN LEVEL

3-4" CONCRETE

2" SAND

PEA GRAVEL

5 MIL POLYETHYLENE VAPOR BARRIER

COMPACTED SOIL

truck arrives. Plan in advance how to get the truck as close as possible to the pour area. Determine where you want to start pouring. Advance planning will save money. Most wet-mix distributors charge waiting fees for any time spent on the site after a 5- or 10-minute period.

If you are mixing your own concrete, start with a test batch. The amount of moisture in the sand and gravel will determine how much water you must add. In your wheelbarrow or mixer, blend 2-1/2 shovels of sand, 1 shovel of cement and 3 shovels of gravel. Mix thoroughly. Make a depression in the middle of the mixture and add 3 gallons of water while mixing with a hoe or shovel. Mix from the inside of the pile out, using a rolling or folding motion.

To test your batch, work it with a trowel. The mixture should slide easily off the trowel, but not run off in a soupy mixture. It should flatten to a smooth finish when smoothed with the trowel. The exposed gravel should be evenly covered with sand and cement paste. If your test batch is too stiff, reduce the amount of sand and gravel. If it's too soupy, add sand and gravel. Keep the water volume constant. Keep notes as you go along so your next batch will be the proper consistency.

If you're using a motorized mixer, first shovel in the gravel or crushed stone and add half the water. Start the mixer and add sand, cement and the rest of the water. Run the mixer for four or five minutes.

After mixing and pouring the concrete, you must level and surface it before it sets. To complete the patio, follow these steps:

1. Shovel and rake wet concrete evenly throughout the form, starting at one end. With a steel rake, carefully rake the concrete level. Work gently. Too much agitation causes the gravel to settle. Level concrete about 1/2 inch above the form to allow for settling.

2. Use a 3- or 4-foot length of 2x4 or 2x6 to *screed* the surface, as shown in the photo sequence at right. For narrow patios and walkways use a board long enough to be placed across the tops of the forms. Pull the screed across the surface with a back-and-forth motion to level the concrete. Make sure you fill all depressions and low areas.

3. Further smooth the surface by *floating* it. Use a bull float for large areas, a darby for small areas. Push the float outward from you in a zig-zag motion. Keep the front edge of the float raised slightly. Then pull it back flat against the surface.

4. Now you are ready to finish the concrete. First, it must be edged with the tip of a pointed mason's trowel. Cut between concrete and form, about an inch deep. Then scrape all concrete off the form.

5. After the water sheen on the concrete surface has evaporated and concrete has begun to set, work around forms with the edger. Don't allow concrete to pile up on leading edge of the

edger. Final edging can be done later, after the final surface finishing has been completed. Use a jointer to make control joints as described on page 109.

6. When you've done the initial edging and jointing, float the surface again, using a smaller floating tool. For a rough to semismooth finish, use a wooden float. For a smooth finish, use a metal float. Use a metal float for air-entrained concrete.

7. After floating, wait 1/2 hour before final finishing. Then use the appropriate finish tool for the surface you want. For a semismooth finish you can use the metal or wood float again. For a hard, slick finish, use a rectangular steel trowel. Move the trowel in sweeping arcs, flat against the surface. For a rough-textured or patterned finish, sweep the surface with an ordinary broom. Broom finishes provide an excellent non-slip walking surface around swimming pools and other wet areas.

8. Allow the concrete to cure. It must be kept constantly moist for five to seven days, depending on air temperature. Cover the concrete with plastic sheeting weighted at the edges or with well-washed burlap. Keep burlap damp until curing is complete. *Ponding* is an option for small areas, such as a patio boarded by masonry walls or concrete edgings. In ponding, the concrete surface is covered with a layer of water 1/2 to 1 inch deep until curing is complete. Commercially produced agents are available for quick-curing concrete.

Installing A Concrete Patio

Pour concrete into the forms and spread with a rake or shovel. Pour concrete about an inch above the forms to allow for settling.

Use a 3- to 4-foot 2x4 to level, or screed, the concrete. Pull the 2x4 screed toward you with a back-and-forth motion.

A special concrete tamper called a *jitterbug* is used to level and compact wet concrete. Used for large patio areas, the jitterbug is operated like a tamper to keep aggregates in the concrete from settling to the bottom of the pour. You can buy this tool at a masonry tool supplier or rent one at a tool rental outlet.

Smooth the leveled concrete with a bull float.

If you're using a concrete-curing agent, you can apply it at this point or wait until the concrete is finished.

Use an edging tool to round concrete edges and separate them from the form.

Use a jointer to make control joints. Space joints no farther than 10 feet apart.

Smooth the surface with a float. Move the float in sweeping arcs. For a semismooth surface, use a wood float. For a smooth surface, use a metal float. Cure the concrete as described on page 110.

COLORING CONCRETE

Your concrete patio needn't be gray. It can easily be tinted before, during or after pouring. Many color pigments are available. Earth tones are best suited to natural garden surroundings.

Pigments—If you add pigment to concrete mix, follow package instructions. The amount of pigment usually should not exceed 10% of the cement's weight. To assure even coloring, use precisely the same proportion in every batch. For brighter color, make a mixture of pigment, Portland cement and water. Apply the tinted mixture 1 inch thick over the rough-finished slab and complete the patio surfacing.

Dust-on color is applied while the concrete is being floated. Carry the concrete-laying process through the wood-floating stage. Spread 2/3 of the required amount of pigment evenly over the slab. Then float the surface, working the pigment evenly into it. Add the rest of the pigment dust. Float once more, and finish the surface by troweling lightly. Repeat edging and joint-grooving processes after each application of color.

Surface Coatings—There are a number of paints, waxes and stains suitable for concrete surfaces. Several manufacturers offer latex and oil-base masonry paints. Latex is easier to work with and equals oil-base paint in durability.

Stains are inexpensive and durable finishes. Semitransparent wood stains work well on concrete surfaces. Stain may be applied after concrete has cured for two months or longer. Wire-brush the surface, using trisodium phosphate and warm water on greasy or dirty areas. Rinse thoroughly and let dry. Then brush or roll on stain. For a darker surface apply a second coat of stain.

Pigmented concrete waxes can be applied to semismooth concrete surfaces. They don't work well on hard, slick surfaces. The low porosity of the surface does not allow good bonding and penetration. Before applying wax, clean the surface with a wire brush and a 10% muriatic-acid solution. Rinse thoroughly. Waxes can be applied with a sponge, then buffed with a rag after they dry. Waxes are not as durable and long-lasting as paints or stains.

KOOL DECK

This is a trade name for a masonry surfacing material applied directly over freshly poured concrete. A commercially prepared mixture is combined with sand and white cement to form a concretelike material. The mixture is splashed with a brush over the finished concrete slab, then troweled to create the texture shown in the photo below left.

Kool Deck has several advantages over concrete. As the name implies, the surface stays much cooler than ordinary concrete. This makes it popular in hot climates. It makes a durable, nonslip surface around pools and other wet areas. Kool Deck is stain and mold resistant and is available in several pastel colors. Applied correctly, it forms a permanent bond with the concrete base and will not crack due to expansion and contraction.

Kool Deck is usually applied by concrete contractors or swimming-pool contractors. Do-it-yourself instructions are available from the manufacturer.

Kool Deck is a masonry surfacing material applied directly over freshly poured concrete. The nonslip surface texture is ideal around pools and spas. The surface stays cooler than ordinary concrete.

Green pigment was added to concrete when this patio was poured.

Latex or oil-base masonry paint makes a durable surface covering. Apply paint with a roller or brush. Don't forget to block off the area while paint is drying.

EXPOSED-AGGREGATE CONCRETE

Exposed aggregate is an attractive and popular concrete-surface variation. It can provide texture and color to add interest to your patio. Exposed-aggregate surfaces look more natural than finished concrete. They also reduce glare. To soften large expanses of concrete, alternate sections of exposed-aggregate and smooth-finished surfaces.

Exposed-aggregate surfaces require more time and effort to make than finished-concrete surfaces.

There are two basic ways of creating this surface. One is by mixing the desired aggregate, usually decorative pebbles, into the concrete before it's poured. The pebbles take the place of the gravel. The other is by seeding the aggregate into the concrete after it's been poured and surfaced, before it sets.

Seeded Aggregate—To seed aggregate, spread pebbles over the concrete just after it has been leveled and rough-surfaced with the bull float. Use a float to press aggregate just below the concrete surface. Then finish as described in the following.

Premixed Aggregate—If you mix the aggregate into concrete, follow the usual procedures for pouring and finishing concrete. Be careful while leveling and floating the concrete. Too much agitation will cause the pebbles to sink too deep to be exposed.

When the concrete has set enough to support your weight, you can start exposing the aggregate. This is usually about 1 to 2 hours after pouring. Put down a plywood kneeling board to distribute your weight. Hose off the surface, using a fine mist. Then use a soft-bristle brush or broom to gently brush away surface paste to expose the aggregate. If any pebbles dislodge, the concrete isn't yet hard enough for brushing.

Allow surface water to evaporate. Then repeat the gentle brushing and hosing, until about half of each pebble is exposed. Finish with a fine spray to wash the aggregate. Finally, cure the concrete as described in Step 8 on page 110.

Exposed-aggregate patios require the same site preparation as finished-concrete patios. Here, wooden forms establish an interesting pattern of squares and rectangles to break up a large patio area.

Aggregate is exposed by spraying the surface with water and brushing with a broom.

Large exposed-aggregate slabs separated by a ground cover of baby's tears form a cool, inviting entry walk.

Adobe

Adobe is a practical and beautiful construction material traditionally used in the American Southwest. Its earth tones harmonize with the natural surroundings. It is suggestive of native American and Spanish architectural styles.

Technological advances have made modern adobe materials more durable than traditional adobe paving. Adobe is an appropriate choice for building a sturdy and inviting Southwestern-style patio.

In some parts of the Southwest and West Coast, adobe may cost less than brick or concrete. Elsewhere, higher delivery costs make it more expensive where available.

Advantages

● Warm, earthy color, complements Southwestern architecture.
● Inexpensive in areas where readily available.
● Easy to install due to uniformity of shape and size.

Disadvantages

● May not be available in all areas.
● Ordinary, untreated adobe is easily worn down by weathering and heavy foot traffic. Most modern adobe materials are manufactured to resist these elements.

SIZES

Adobe bricks for patio paving are stronger than those used for wall construction. Adobe patio bricks are usually available in two sizes: 12x12x2-1/2" and 6x12x2-1/2". To pave a 100-square-foot patio area, use 100 of the larger or 200 of the smaller bricks. Remember to buy a few extras for future replacement. Without spares, you may have a hard time matching the exact size, shape and color of the bricks you just bought.

LAYING ADOBE BRICK

Manufacturers recommend butting the bricks tightly for stability on a 2-inch bed of tamped, screeded sand. Because size and shape of individual units may vary slightly, the joints won't be as tight as those of ordinary brick. If laying bricks on sand, you'll need more sand for joints than with ordinary brick: about 11 cubic feet of sand for a 100-square-foot area.

Installation procedures are the same as for laying bricks on a sand base. See page 104. Extremely narrow, sand-filled joints between bricks permit drainage. Treated adobe bricks are waterproof.

You can lay adobe bricks with 1-inch, soil-filled joints. This method allows you to plant in the joints for a softer effect. The bricks may not be as stable as if they were tightly butted, but they will be more stable than ordinary bricks set in the same manner.

If possible, avoid filling the joints with mortar. Mortar prevents drainage through the joints. Salts in the mortar eventually damage adobe bricks.

Adobe brick is best laid in the simple jack-on-jack pattern. See page 105. This method minimizes brick cutting. If you do need to cut an adobe brick, use the same procedure as for ordinary bricks, described on page 106.

Adobe is one of the most stylistic patio materials. Its warm earth tones complement Southwestern architecture.

Concrete Paving Blocks

Cast-concrete paving blocks are sold in many sizes, shapes, colors and textures. They are cheaper than stone or terra cotta blocks. The least-expensive alternative is homemade concrete paving blocks.

Advantages
• Available in many sizes, shapes, and colors. Many simulate other materials, such as brick, stone, or tile.
• Generally less expensive than brick, tile, or paving stones.
• Durable.

Disadvantages
• Most types available have a rough-textured concrete look.
• As with other individual paving units, a concrete block surface makes some activities, such as dancing or riding children's toys, difficult.

Concrete paving blocks, called *cap blocks,* make an excellent inexpensive patio surface. Block sizes are 6x12x2", 12x12x2", and 24x24x2". These can be laid over a sand bed or in mortar over a concrete surface. Follow instructions for laying brick, starting on page 104.

Homemade paving blocks can be poured in a ground mold, then used in place or moved. For paving a large patio, build a grid of forms to cast the blocks in place. Use 2x4s to build forms in a regular pattern or in a combination of sizes and shapes. Taper the sides of the forms slightly toward the bottom. Wet the forms thoroughly, then cover sides with a light coat of clean motor oil. Place the grid on properly graded, packed soil. Fill the forms with concrete. Next, use a short length of 2x4 to level or *strike off* the block surfaces. Then finish the surfaces with a wood float.

When the concrete has hardened enough to support itself, use a pointed trowel to edge around the blocks and gently lift off the forms. Reposition the blocks to the desired joint width. Fill the joints with sand or mortar exactly as you would for brick.

You can make exposed-aggregate or stone-and-pebble surfaces if you want. Follow the steps outlined on page 113.

Interlocking pavers make a stable patio surface. Individual pavers are less likely to shift out of position. This pattern gives the patio a quilted look. Interlocking pavers come in many patterns and colors.

Black is a traditional color used in oriental gardens. These smooth, black concrete steppingstones were custom-made for this landscape.

These 4-inch-thick concrete bricks can be used for patio, walk or wall construction.

Ceramic Tile

Use ceramic tile to create a precise, formal patio. There are literally thousands of color combinations and hundreds of textures.

Advantages
- Makes a formal, well-tailored patio.
- Provides a beautiful transition between indoor and outdoor areas.
- Available in a variety of sizes, shapes, colors, and surface textures.
- Provides a durable surface that is easy to maintain.

Disadvantages
- Expensive.
- Must be laid on a perfectly smooth, level surface.
- Glazed tiles can be slippery and cause glare.

KINDS OF TILE

Tile is categorized in several ways. First, there is glazed tile and unglazed tile. Glazed tiles have a glasslike finish and are generally used indoors. The glazed surface is slippery when wet. Glazed tiles are sometimes used for indoor-outdoor patios, if the entire patio surface is protected from the weather. Unglazed tiles are recommended for outdoor use.

Indoor tiles are generally smaller, thinner tiles. Outdoor tiles are larger, thicker tiles made to withstand outdoor conditions.

Outdoor tiles are further broken down into two groups—patio tile and quarry tile. Patio tiles come in brick-red and earth tones, with irregularities in shape and texture. Quarry tiles are more regular in shape and lend a cleaner, more-precise look to the patio.

When choosing tile for outdoor use, consider the location in which it will be laid. Glazed tiles are less likely to stain and are easier to clean than unglazed tiles. Glazed tiles are excellent for areas subject to dirt and grease, such as barbecue or eating areas.

Because glazed tiles are slippery, choose ones with a textured surface. Unglazed, rough-textured tiles are preferable for use around swimming pools and for patios fully exposed to the elements. Some tiles are intentionally designed to be slip-proof.

All outdoor tile, like indoor tile, is durable and easy to maintain. Tile is one of the most-expensive paving materials you can buy.

Heavy tiles can be can be laid on a

Outdoor patio tiles are most commonly available in the size and color shown here—12x12" and brick-red. Unglazed tiles usually must be treated with a sealer to help resist the weather. Smooth tiles are slippery when wet.

sand bed or directly on well-drained soil. They can also be laid on mortar, as all thinner tiles should be. All tiles must be laid on a smooth, level surface. Slight variations in grade will show up in the finished surface. Existing or new concrete slabs and wooden decks make especially suitable surfaces for laying tile.

You can butt tiles together and leave the joints open, but the preferred method is to leave 3/8- to 5/8-inch joints and fill them with mortar. Mortared joints make a more-stable installation and give the surface a finished appearance.

To lay tiles on mortar, first lay a 3-inch bed of compacted sand within 2x4 forms. Spread a 1-inch layer of mortar over the sand and work it with a float or trowel until the mortar bed is smooth and level. If you're working on a concrete slab, first wash the slab with diluted muriatic acid. Rinse well before applying mortar.

Apply the mortar in sections no larger than 20 square feet so the mortar won't harden before you set the tiles. Soak the tiles in water for 5 to 10 minutes. Remove the tiles, let

dry a few minutes, then set them in the mortar bed. Use 3/8- to 5/8-inch joints. Actual joint size should complement the size of the tile you've chosen. Tap tiles gently into place. Allow them to set for at least 24 hours. Then mortar the joints, using a mortar mixture of 1 part cement, 3 parts sand, and enough water to make a soft mix. Keep the tiles clean as you go. Use a 1/2-inch dowel to smooth the joints. Allow the mortar to set for three to four days before walking on the patio surface.

Tiles can be laid over a wood deck. Be sure the deck is strong enough to support 20 pounds *live load* per square foot. Live loads and dead loads are discussed on page 125.

First lay 30-pound asphalt felt over the deck surface and nail it in place with galvanized roofing nails. Cover the felt with loosely nailed 3/4-inch stucco wire. Chicken wire will also work. Cover the wire with a 1-inch mortar bed dusted with a layer of cement. Set the tiles in the mortar bed and mortar the joints as described above. Use a mortar mixture of 1 part sand to 5 parts cement.

Flagstone

Flagstones are strikingly handsome in almost any patio setting. They're a natural component of the earth itself. They stand up under the hardest wear.

Advantages
- Natural in appearance.
- Available in many textures and colors.
- Can be laid directly on stable soil.

Disadvantages
- Very expensive.
- Can be slippery when wet.
- Pieces come in irregular shapes, making it difficult to maintain an overall pattern.
- Too much flagstone can be confusing to the eye because of the irregular shapes.

KINDS AND SIZES

Slates and sandstones are most often used for this versatile, natural patio-paving material. Unfortunately, they're the most-expensive paving material, though their beautiful appearance may justify the cost.

You may want to work with irregular shapes, creating a random pattern and filling between stones. However, random pieces can be frustrating to lay, leading to unsightly results. It may be better to stick to pieces varying from 1 foot wide in smaller dimension to 4 feet wide in larger dimension.

Whether you use regular or irregular stones, thickness should be uniform. The minimum thickness for use in mild climates is 3/4 inch. In areas with severe winters, use flagstones 3 inches thick or more. Two inches is adequate in areas with average climates.

CUTTING THE STONE

If you find it necessary to cut flagstones, it's easiest to take them to a stonecutter to be sawed. If you do the cutting, use a brick set and a heavy hammer. Mark the cutting line on both sides of the stone and then score along the lines about 1/8 inch deep. Score by gently tapping the brick set with the hammer.

Center the score line over the outside edge of a 2x4 or 2x6. Put the brick set in the center of the line and tap sharply with the hammer.

LAYING FLAGSTONE

For maximum durability, lay flagstones in a bed of mortar on a 3-inch concrete slab. Follow the same procedures given for laying bricks over concrete on page 106. Make joints between stones 1 to 2 inches wide. The concrete slab and mortar bed make it possible to use very thin flagstones.

No foundations are necessary for laying flagstones in stable, well-drained soil such as sand or loose loam. For poorly drained soil, use a 2-inch bed of tamped and leveled sand. A deeper bed may be necessary in soil subject to frost heaving.

Lay flagstones as tightly as possible on soil. This helps prevent weeds and grasses from growing between them and ensures stability. For a more finished appearance, mortar the joints. Another alternative is to lay stones 3 or more inches apart, then plant grass or a low ground cover in the gaps.

This flagstone patio has a natural, yet elegant appearance. Mortaring joints between stones makes a more permanent, formal-looking surface.

Stones and Pebbles

Stones and pebbles are among the most useful and versatile of patio paving materials. You can pave an area with cobblestones to create a rustic European effect. You can lay smooth river rocks or pebbles to create an oriental effect.

Advantages
- Widely available in every size, shape and color.
- Natural in appearance.
- Easy to install.

Disadvantages
- Can be slippery when wet.
- Difficult to match sizes and shapes when installing.
- Can be rough on bare feet.

Laying pebbles and stones is easy. Just remember that if they are laid at different depths, the resulting surface will be uneven and difficult to walk on.

Larger stones can be set in soil or mud, preferably clay. Be sure they are securely set and safe to walk on. Edgings may be used if needed. A board is useful to press smaller stones and pebbles into the ground to create a level surface.

You can set stones and pebbles in mortar over concrete. See the section on laying bricks in mortar on page 106. Sections of stones or pebbles help break up a monotonous expanse of concrete. You can set stones and pebbles in new concrete itself. Just use the same method as for seeding aggregate. The possibilities are limited only by your budget, time and creativity.

Steppingstones made of pebbles embedded in concrete provide a smooth walking surface along crushed-granite path.

Different color pebbles are meticulously set in concrete to form a geometric design.

White creek stones set in concrete complement birch trees and informal landscape in background. Railroad ties serve as permanent forms for steps.

Wood

Though often thought of as a deck material, wood can make a beautifully warm pavement for your patio. You can use it alone or in combination with other paving materials. Decay-resistant woods, such as redwood, red cedar, cypress and pressure-treated woods are suitable for on or below-ground use.

Advantages
- Natural appearance.
- Easy to install. Available in many sizes and shapes.

Disadvantages
- Unless pressure-treated with preservative, even decay-resistant species have a short life compared to masonry patio surfaces.

WOOD USED FOR PATIOS

Several forms of wood paving materials are available.

Wood rounds, or cross sections of trees, make handsome informal paving units. Redwood and red cedar make attractive rounds. Because the lighter sapwood in the round is not decay resistant, the entire round should be soaked in a wood preservative if it is to be set in concrete or mortar. Rounds set in the ground are easily replaced should they decay. For tips on treating wood in preservatives, see page 127.

Lay rounds on a leveled, well-tamped sand base. You can fill between them with sand or bark chips.

End-grain wood blocks make a durable patio surface. You can cut your own from large timbers. Lay them close together on a 1- or 2-inch bed of tamped sand. Sweep sand over them to fill the narrow but irregular joints. If you use pressure-treated material, the cut ends must be treated with preservative.

Railroad ties can be sunk flush with the ground or laid on top of it. They are expensive if used over a large patio area, but they make excellent walkways and steps. They're often used in combination with other paving materials. Railroad ties have already been treated with a preservative, usually creosote.

If ground moisture is a problem, spread a sheet of polyethylene under the wood before laying it. Then lay the bed of sand for rounds or blocks.

If your soil is unusually hard or if winter ground heaving is a problem, prepare the site. Lay a base of 3-4 inches crushed stone or gravel and 2 inches of sand beneath the wood pavers.

Wood in contact with the ground should be treated with preservative. Buy pressure-treated wood or treat the wood yourself, using the following method.

Soak wood rounds or blocks for several days in preservative. Allow wood to dry before it's laid. Because some preservatives are highly toxic, handle them with care. Follow application instructions on the container label. For more information on applying wood preservatives, see page 127.

4x8" wood blocks laid on end make a warm surface for patios or walkways.

Loose Aggregates

Loose aggregates are materials such as wood chips, gravel and crushed granite. Though they're usually not considered a permanent paving for patios, they do have a number of uses in the yard.

Loose aggregates offer a number of possibilities. They can be used to cover paths, children's play areas, or party spill-over areas adjacent to permanent patios. Use them under large trees to allow air and water to reach tree roots.

Loose aggregates are relatively inexpensive. They are easy to apply and come in many colors and textures. They are easy to replace if you want a more permanent surface later. After a few year's wear, some loose aggregates provide a sound foundation for a permanent patio.

Various types of loose aggregates are summarized in the chart below. Availability of some loose aggregates depends on where you live. You may discover possibilities not listed on the chart. Locally quarried rock, walnut shells, pecan hulls or fruit pomace from local crops can be inexpensive alternatives.

Loose aggregates create some problems not encountered in permanent patios. Weed growth is one. It can be controlled by laying down several layers of black polyethylene plastic under the aggregate. Punch holes in the plastic for drainage. An alternative is to spray the area with weed-killer.

To keep loose aggregates in place, use a strong edging as described on page 122. With heavy aggregates, such as crushed gravel or rock, put down a single layer of aggregate. Wet the aggregate thoroughly and roll it out with a water-filled sod roller. This will help keep the aggregate in place.

Loose aggregates need to be replenished periodically. How often depends on the type of aggregate. Organic materials such as wood chips decay rapidly and must be replenished most often. On paths and other walking surfaces, hard aggregates such as gravel usually need replenishing every four or five years.

Try to keep aggregates relatively clean. Dirt and other organic matter mixed in the aggregate encourages weed growth.

Smooth, mixed-aggregate rock transforms this winding path into a dry stream bed.

AGGREGATE CHARACTERISTICS

Material	Coverage 100 sq.ft. 2" deep	Advantages	Disadvantages	Comments
Wood Chips	1 cu. yd.	(1)	(2)	(3)
Gravel	2/3 cu. yd.	(4)	(5)	(6)
Redrock	3/4 cu. yd.	(7)	(8)	(9)
Crushed Granite	3/4 cu. yd.	(10)	(8)	(12)
Crushed Brick	1 cu. yd.	(13)	(14)	(15)

ADVANTAGES
1. Soft textured, natural looking.
4. Blends well with other paving materials. Dries quickly, easy to clean.
7. Forms a compact, clean surface when rolled.
10. Forms a compact, easy-to-clean surface.
13. Good match for brick and tile.

DISADVANTAGES
2. Decays rapidly. Tends to spread into areas where it's unwanted. Harbors ticks in some situations.
5. Can be hard on bare feet.
8. Can be hard on bare feet. Gradually crushes and breaks down with time.
14. Expensive. Not meant for heavy traffic or bare feet. Breaks down quickly.

COMMENTS
3. Locally available in many colors and textures. Good for children's play areas.
6. Available in many sizes, colors and shapes.
9. Can be used as a base for a permanent patio after a few years wear. May be locally available under other names.
12. Similar to redrock but lasts longer. Natural brown and gray colors.
15. Can be used as a base for a permanent patio. Choose colors carefully. Red can be very bright.

Shredded bark, left, and chunk bark, right, are kept in place by pressure-treated timbers to form low steps up gentle slopes.

Decorative gravel and concrete steppingstones lead to a low-level wood deck. Gravel is held in place by railroad-tie borders and hardy, low-growing junipers.

Loose-aggregate patio can be a temporary problem solver. This gravel patio makes an attractive ground cover. If the owner decides to install a more permanent patio, the gravel will make an excellent base.

Edgings

Most patios use one or more types of edging. An edging provides a decorative break between one surface and another. An example is a wooden beam between the patio and lawn. Forms and construction headers may serve as permanent edgings. You can use decay-resistant wood or almost any type of masonry for edgings.

To set a wood edging, run a string to mark the exact outline of the patio. For flush edgings, use a square-point shovel to dig a trench slightly deeper than the edging thickness. Drive stakes no farther than 4 feet apart and nail the edging to it. Cut stake tops level with the top side of the edging, angling the cut away from the edging board. Backfill the trench and compact the soil to help support the edging.

A masonry edging can be a single row of bricks, called a *soldier course,* or it can be concrete poured into a narrow form. For durability, a brick edging can be set in a mortar base of 1 part cement and 3 parts sand. Bricks can be laid flat, on edge, on end, or diagonally.

Other materials, such as stones or paving blocks, make attractive, durable edgings. Browse through home improvement centers and masonry suppliers to find unusual, functional edgings.

Railroad ties hold bricks in place. Flush edging separates brick path from loose-aggregate seating area. Edging in foreground creates a raised planter bed.

Exposed-aggregate concrete edging provides contrast for finished concrete patio.

Loose-set bricks make an informal edging for this gravel path.

Bricks on edge, called a *soldier course*, make the edging for this brick path. The edging helps contain soil in adjoining planter beds.

Deck Builder's Guide

This chapter on deck building presents step-by-step instructions and suggests many alternative building methods. It is liberally illustrated to help you through each step of construction. If you need additional advice on any phase of construction, seek expert help *before* you start building. This help may save you a good deal of time and money.

Building a deck can be a weekend-long or summer-long project. It depends on the type of deck you build and the amount of time you have to build it. No matter how long it takes, building a deck requires careful planning. If you have not already made plans for your deck, see pages 9-30 in the chapter, "Planning Your Patio Or Deck." The chapter will tell you how to design a deck and surroundings to suit your outdoor-living requirements. It will tell you what is required to get a building permit.

WORKING DRAWINGS

In addition to an overall landscape plan, you'll need *working drawings*. These are scale drawings that show the overall dimensions of the deck. The drawings show the relationship between structural elements, such as posts, beams, joists and deck boards. Include lumber sizes on your drawings.

Make your working drawings to scale on a 24x36'' sheet of 1/4-inch graph paper. A good scale is 1 square equals 2 inches, or 6 squares equal 1 foot.

Your working drawings should include two top-view drawings, called *plan views*. One plan view shows the locations of footings, posts, beams, joists, ledgers and other deck substructure members. The second view shows the surface pattern of the deck boards—straight, diagonal, parquet or

Left: Decks can easily create usable space in unused portions of the yard. This deck is being built over a rocky slope. No grading or rock removal was necessary.

other design. This drawing shows the size and location of steps, railings, benches and other surface structures.

A second drawing, called an *elevation view*, shows details of all vertical members of the deck. Simple low-level decks may not require an elevation view. Raised decks or multilevel decks require an elevation view, especially if the substructure has diagonal bracing or is in some other way complex. You may want to draw elevation views of railings, overheads or benches.

In order to get a building permit, you may need to supply the building department with working drawings along with your site plan as described on page 28. If the department doesn't require them, the drawings are still helpful for estimating materials and for reference during construction. If a contractor is doing the work, the plans help ensure that you'll have the deck built the way you want it.

Before you start your drawings, study the rest of this chapter. You must have a good understanding of deck materials and construction practices to make accurate drawings of your deck.

DESIGNING A SOUND DECK

A key element in deck design is determining the proper sizes, spans and spacings of the deck members. Use the charts on pages 132-133 to determine the sizes and locations of posts, beams, joists, and deck boards. Consult local codes for structural requirements.

The deck must have a firm foundation. The foundations of most decks consist of *footings* and *piers*. These support the posts, which in turn support the rest of the deck. Most deck foundations need not be extensive. Decks built on steep hillsides, unstable soil, or ones in cold climates subject to severe freezes require more extensive footing and pier systems. These should be designed by an

expert. Consult your local building department for advice on foundation requirements.

Weight is a factor in deck design. The weight of the deck members themselves, called *dead load*, and the weight of everything placed on the deck, called *live load*, help determine lumber size, span and spacing, and foundation requirements. The span and spacing charts on pages 132-133 are based on standard live and dead loads. See footnotes in charts.

If the live load on the deck will be excessive, such as a hot tub full of water or a large masonry barbecue, you'll need a much stronger foundation and substructure for your deck. It's better to build separate foundations for these items and build the deck around them. Your local building department has information on structural requirements for various loads.

Tools And Materials

Wood is the primary deck-building material. Secondary materials include concrete for piers and footings, fasteners such as nails, screws, bolts and metal connectors, and finishes such as paints, stains and clear sealers. You'll need tools for measuring and marking, foundation work and carpentry work.

TOOLS

You don't need an elaborate set of tools to build a deck. The only power tools you'll need are a *portable circular saw* and a *drill*. A *portable jigsaw* comes in handy for cutting curves in the deck surface or for rounding off square edges for ornamental effects.

Measuring and marking tools include a *pencil*, a *50- or 100-foot metal tape measure*, a *framing square*, a *try square* or *combination square*, a *carpenter's level*, a *line level*, *sturdy nylon string*

Most of the tools needed to build a deck are shown here. Not included are a shovel and wheelbarrow for mixing and pouring concrete footings, and painting equipment.

or *twine,* a *plumb bob* and a *chalk line.* Optional measuring tools may include a *10- or 12-foot tape measure* or a *folding rule.*

You'll need a *shovel* to dig footing holes for the deck foundation. A *pick* is useful for breaking up hard or rocky soil. Carpentry tools you'll need include a *hammer,* a *handsaw,* a *nail set,* and a *chisel.* A *wood rasp* works well for cleaning notches in wood for tight-fitting joints.

A *wheelbarrow* is handy for mixing concrete for footings. If you have very many footings to pour, consider renting a *motorized cement mixer.* Specialized concrete-working tools are not necessary for the simple concrete work on deck footings. Don't forget brushes, rollers and other painting equipment for applying finishes.

SELECTING LUMBER

There are two main considerations in selecting lumber for your deck. The first is the *species* of wood you choose. The second is the *quality* of the pieces you select. Both considerations help determine the strength and durability of the deck.

Wood Species—A number of wood species are used for wood decks. Each has its good and bad points. Some woods have greater structural strength, others have better decay resistance. Some have a more-attractive appearance than others. You'll need to choose woods that meet your deck's structural and appearance requirements. Your choice may involve several types of wood.

Availability and price of different species will influence your decision. For instance, don't set your mind on building with red cedar if no lumberyards in your area stock it. You might be able to get it, but a special order would probably be expensive.

Some parts of the deck substructure will require a naturally decay-resistant wood such as redwood, red cedar or cypress, or wood that has been treated with a preservative. For other parts of the deck, decay resistance is less important.

Softwoods are generally used in outdoor building. Softwood lumber comes from trees with needles. *Hardwood* comes from trees with leaves. These terms do not mean that the wood is hard or soft. For example, Douglas fir, a softwood, is much harder than balsa, a hardwood. Softwoods are usually easier to work with than most hardwoods. They're also less expensive and come in a wider variety of lumber sizes.

Most softwoods are grown and milled in the western and southern United States. Southern pine and bald cypress are generally more available in the South and the East. Douglas fir, Western hemlock, pines, red cedar and redwood predominate in the West. The Midwest is an in-between region where availability of each species is unpredictable.

The chart below shows the strength, working characteristics and relative hardness of softwood species commonly used for decks. If you have a choice between several woods, use

WOOD CHARACTERISTICS[1]

KIND OF WOOD	WORKING AND BEHAVIOR CHARACTERISTICS							STRENGTH PROPERTIES			
	Hardness	Freedom from Warping	Ease in Working	Paint Holding	Nail Holding	Decay Resistance of Heartwood	Proportion of Heartwood	Bending Strength	Stiffness	Strength as a Post	Freedom from Pitch
Western red cedar	C	A	A	A	C	A	A	C	C	B	A
Cypress	B	B	B	A	B	A	B	B	B	B	A
Douglas-fir, larch	B	B	B-C	C	A	B	A	A	A	A	B
Hemlock, white fir[2]	B-C	B	B	C	C	C	C	B	A	B	A
Soft pines[3]	C	A	A	A	C	C	B	C	C	C	B
Southern pine	B	B	B	C	A	B	C	A	A	A	C
Poplar	C	A	B	A	B	C	B	B	B	B	A
Redwood	B	A	B	A	B	A	A	B	B	A	A
Spruce	C	A-B	B	B	B	C	C	B	B	B	A

[1]A—among the woods relatively high in the particular respect listed; B—among woods intermediate in that respect; C—among woods relatively low in that respect. Letters do not apply to lumber grades.
[2]In the West, often graded as hem-fir; also refers to Eastern hemlocks.
[3]Includes the Western and Northeastern pines.

the chart to help you make your decision.

Decay-resistant Wood—Wood will eventually deteriorate under adverse conditions. It won't decay if kept dry—less than 20% moisture content. Because decks are usually not protected by a roof, they are subject to continual wetting and drying. Substructure members—posts, beams, and joists—are especially vulnerable to decay.

To assure long deck life, use a naturally decay-resistant species or a wood that has been treated with a preservative. Redwood, red cedar and cypress are decay-resistant species, but only the *heartwood* of these trees resists decay. The heartwood is the darker-colored wood cut from the center of the tree. The lighter-colored sapwood has little more decay resistance than non-resistant species such as pine or fir.

Whatever wood you choose, use construction techniques that do not trap moisture. Some joints are more prone to trapping moisture than others. To prevent moisture buildup underneath the deck, provide adequate air circulation to all substructure members.

Pressure-treated lumber is wood that has been impregnated with a preservative by a pressure process. This process forces preservative deep into the wood cells. Pressure treatment is much more effective than simply soaking the wood in a preservative.

You may find pressure-treated wood identified as *Outdoor Wood* or *All-Weather Wood*. You'll be able to identify it by the stamp of the American Wood Preservers' Bureau combined with other identifying marks.

There are several preservatives used in pressure treatment. Two of these, *chromated copper arsenate* (CCA) or *ammoniacal copper arsenate* (ACA), are classified as *waterborne salts*. Waterborne-salt preservatives are initially dissolved in water when applied to the wood. Once applied, they do not leach out of the wood under wet conditions. ACA and CCA are clean and odorless, but they tinge the wood green, blue green, light yellow or brown. Surfaces treated with ACA and CCA can be stained or painted, or left unfinished. Once applied, these preservatives are non-toxic to plants and animals.

Pentachlorophenol, commonly called *penta,* is another popular preservative

for pressure-treatment. Wood-surface color and paintability depend on the solvent base used with penta, as described below.

Pressure-treated wood falls into several categories, depending on the amount of preservative used and the depth of penetration. These two factors determine the wood's *retention* of preservative. Treated wood marked LP-2 is treated to a minimum retention of 0.25 pounds of preservative per cubic foot of wood. This wood is recommended for above-ground use. Wood marked LP-22 is treated to a minimum of 0.40 pounds per cubic foot and can be used on or below the ground.

Some pressure-treated lumber has marks on the surface made by small blades, called *incisors*. This permits more complete penetration of the preservative. Lumber with incisor marks can be used on deck substructure members hidden from view.

Preservatives used for pressure treatment do not leach out of the wood. The Forest Products Laboratory of the U.S. Department of Agriculture has determined that properly pressure-treated wood will last a minimum of 50 years, if used correctly. This wood lasts about five times as long as naturally decay-resistant woods.

The cut ends of pressure-treated lumber pieces should be soaked with preservative before they are put in place.

Preservative Treatment—You may find that pressure-treated wood or naturally decay-resistant wood is not readily available in your area. If this is the case, it might be less expensive to treat a non-resistant species with preservative. Treating the wood yourself is not ideal, but it may be necessary if no other options are available. Do-it-yourself treatment will extend the life of the wood by 5 to 10 years. As mentioned, pressure-treated woods generally last 50 years or more.

Soaking the wood overnight in preservative is more effective than simply brushing the preservative on or dipping the wood in it. Treating the wood yourself does not offer the protection pressure-treated wood does.

Here is the best way to apply preservative to your lumber. Make a trough from 4x4s or timbers as shown in the photo on page 128. Line the trough with plastic sheeting. Fill the plastic-lined trough with preservative and

If you treat wood yourself, soak it in preservative at least 24 hours for adequate penetration. The beam above was soaked for 48 hours. Note depth of penetration.

soak the lumber in it.

There are two common preservatives available for the do-it-yourselfer. *Pentachlorophenol*, or *penta*, is a water-repellent preservative. Penta suspended in a light oil base will not stain the wood and can be painted over. Penta suspended in a heavy oil base will stain the wood brown. Read the container label to see if the penta you buy can be painted over.

Penta is highly toxic and must be handled with care. Carefully read and follow label warnings and application instructions. Do not use penta near plantings.

Copper naphthenate is another popular preservative. It is useful inside planters because it is non-toxic to plants. Copper naphthenate stains wood green, but it can be painted over.

Use a paint roller to apply wood preservatives to the deck surface.

Make a temporary trough from a polyethylene sheet and 4x4s to soak long boards in preservative.

Use a 30- to 50-gallon drum to soak post ends in preservative. The treated end of the post is set on the pier or footing. Use gloves with toxic preservatives such as penta.

Wood Defects

KNOTS

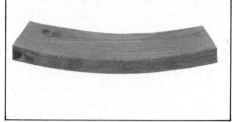

BOW

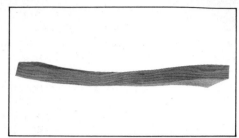

TWIST

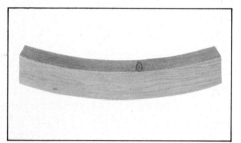

CROOK

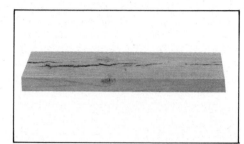

SHAKE

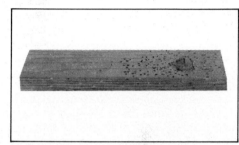

WHITE SPECK AND HONEYCOMB

DECAY

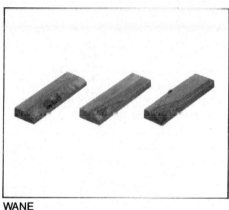

WANE

SPLIT

Lumber Defects—Wood's natural characteristics develop as the tree grows. When the tree is milled into lumber, these characteristics result in defects that affect lumber's strength and appearance. Lumber defects are also caused by errors in milling or damage in shipping.

Become familiar with wood characteristics and defects and how they affect lumber strength and appearance. If the defects are relatively minor, you can often save money by picking through lower-grade lumber piles for usable pieces.

Common wood defects and growth characteristics are:

Knots—these may be sound, loose or missing, leaving a knothole. Knots may surface on one or more faces of a board. Sound knots do not affect lumber strength—loose knots do.

Wane—the presence of bark or lack of wood on the edge or corner of a piece of lumber. This is mainly a visual defect.

Shake—a lengthwise separation of wood that usually occurs between or through the annual growth rings.

Checks—separations of the wood fibers which normally occur across or through the annual growth rings. Checks usually develop as cut lumber dries.

Splits—similar to checks, except separations of wood fibers extend completely through a piece, usually at the ends. Boards with shake, checks, or splits can be used if these defects can be economically cut out.

Bow—a deviation from a flat plane of the wide face of a board from end to end.

Crook—a deviation from a flat plane of the narrow face of a board from end to end.

Twist—a deviation from the flat planes of all four faces by a spiraling or twisting action. This is usually caused by seasoning.

Cup—a deviation from a flat plane, edge to edge. Check for cup, bow, crook and twist by sighting down the end of the board. If minor, these deviations can be corrected when the board is nailed.

White Speck or **Honeycomb**—caused by a fungus in the living tree. White speck is small white pits or spots. Honeycomb is similar, but its pits are larger or deeper. Neither causes further decay unless the wood is used under wet conditions.

Decay—a disintegration of the wood due to wood-destroying fungi. These fungi require a wet environment in which to grow.

Lumber Cuts—Some lumber cuts from a tree are better than others. The cuts are designated by grades for different uses.

In *flat-grain cuts*, the grain is parallel

to the face of the board. In *cross-grain* cuts, the grain angle is up to 45 degrees across the face. Both cuts are generally used for construction or deck boards. *End-grain lumber*, in which the grain is perpendicular to the width of the board, is a higher grade. End grain lumber is used primarily for finish work.

On flat-grain and cross-grain cuts, the surface of the board that was closest to the outside of the tree is called the *bark side* of the board. You can tell the bark side by looking at the growth rings on the end of the board. To prevent deck boards from cupping, position them bark-side up.

Lumber Surfaces—Lumber is classified as *rough* or *dressed*. Rough lumber has not been dressed or surfaced at the mill. It shows saw marks on all four surfaces. Dressed, or surfaced lumber has been run through a planer or sander to attain a smooth surface and a uniform size. It may be surfaced on one side (S1S), two sides (S2S), one edge (S1E) or a combination of edges and sides. In the lumberyard you find mostly rough lumber and lumber that has been surfaced on all four sides (S4S).

Worked Lumber is material such as molding that has been shaped in addition to being surfaced. Consider using worked lumber for ornamental details on your deck such as posts for railings or moldings for benches and planters.

Moisture Content—As you choose your deck lumber, you will notice the grade stamp includes the designation **S-DRY** or **S-GRN**. *Dry lumber* has been dried or *seasoned* at the mill to a moisture content of 19% or less. *Green lumber* may look the same because it has been sawed and surfaced, but was not dried at the mill. By the time you buy green lumber it may have dried naturally to approximately the same moisture content as dry lumber. An easy way to tell green lumber from dry is by weight. Green lumber is heavier.

Some carpenters prefer to work with green lumber because it is easier to nail. Green lumber will shrink once it is nailed in place. This shrinkage must be taken into account when spacing deck boards. Green lumber is more likely to split around nails and warp after the deck is built. If you buy green lumber, allow it to dry at least three weeks before using it.

Green and dry designations also in-dicate lumber size when it comes out of the mill, as explained below.

Lumber Sizes—Familiarity with common lumber sizes and marketing practices will make shopping for deck materials much easier. It will probably save you money.

First, a 2x4 is not 2 inches by 4 inches—2x4 is the *nominal* size—not actual size. A *dry* 2x4 is surfaced at 1-1/2x3-1/2''. A *green* 2x4 leaves the mill slightly larger—1-9/16x3-9/16''. It will eventually *dry down* naturally to match the seasoned size of the dry material. The same goes for other nominal sizes. The chart at right lists nominal and actual sizes for green and dry lumber.

Take actual and nominal sizes into account when designing your deck. Base your overall deck size on actual, not nominal lumber sizes. Use nominal sizes when ordering lumber.

Lumber comes in standard lengths, generally in 2-foot increments. Design your deck to take advantage of standard lumber lengths so you don't waste material.

Lumber Grades—Lumber is graded according to the characteristics and defects described earlier. Grading is different for different wood species. These grades are assigned to lumber when it leaves the mill. Though grading is a good indication of lumber quality, it does not take into account any damage to the lumber during shipping or while in the lumberyard.

It's a good idea to carefully select pieces from the graded piles in the yard. Some lumberyards and home-improvement centers restack lumber according to its actual condition, so it's possible you may find several grades mixed together in one pile.

Lumber grades are important in choosing lumber for beams, joists and decking. Grades help determine the bending strength or stiffness of different species. The stronger the lumber, the greater span it can cross, as indicated in the span and spacing charts on pages 132-133. Local building codes may require certain grades of lumber for different deck members.

You may have to shop around to find the species and grade of lumber required for your deck design. If you can't find it, you'll have to adapt the deck design to accommodate available lumber species and grades.

Lumber is graded by visual inspection at the mill. Grading of *dimension lumber*—pieces nominally 2 to 4

STANDARD LUMBER SIZES		
Nominal (inches)	Dry (inches)	Green (inches)
1	3/4	25/32
2	1-1/2	1-9/16
4	3-1/2	3-9/16
6	5-1/2	5-5/8
8	7-1/4	7-1/2
10	9-1/4	9-1/2
12	11-1/4	11-1/2

These standard lumber sizes are those established by the American Lumber Standards Committee.

inches thick and up to 8 inches wide—is primarily a judgment of end-use strength rather than appearance. Natural characteristics and milling imperfections that affect strength are taken into account. In contrast, grading of *board lumber*—pieces nominally 1 inch thick—is based on appearance rather than strength.

Most lumber is marked with a grade stamp which contains five elements. These elements describe manufacturer, grade, species, moisture content at time of surfacing and certification mark of an approved grading agency. A typical grade stamp is shown in the drawing at right. Grade stamps indicate that pieces in each grade have an established range of strength, appearance properties or both.

Dimension-lumber categories for softwoods are *Light Framing, Structural Light Framing, Studs,* and *Structural Joists and Planks*. Within these categories, the lumber is graded Nos. 1, 2 and 3. See the chart on facing page.

Three basic garden grades of redwood are used in deck construction. *Construction Heart* is used for posts, joists, stringers and other structural parts near the ground. *Construction Common*, which contains sapwood, is used for deck boards, fencing and railings above ground. *Merchantable* grade redwood has sapwood and a few larger knots. It is the most economical of all grades. You can use it for planters, benches, trelliswork and other nonstructural deck elements.

Board lumber is graded differently than dimension lumber. It is used for fences, screens or railings. Generally, boards are nominally 1 inch thick and up to 1 foot wide. While boards are not always grade-stamped, they are graded. You will often find them separated by grades in bins or on shelves. When boards are stamped,

DIMENSION LUMBER GRADES

LIGHT FRAMING 2 to 4" Thick 2 to 4" Wide	CONSTRUCTION STANDARD UTILITY	This category for use where high strength values are NOT required; such as studs, plates, sills, cripples, blocking etc.
STUDS 2 to 4" Thick 2 to 6" Wide 10' and Shorter	STUD	An optional all-purpose grade limited to 10 feet and shorter. Characteristics affecting strength and stiffness values are limited so that the "Stud" grade is suitable for all stud uses, including load-bearing walls.
STRUCTURAL LIGHT FRAMING 2 to 4" Thick 2 to 4" Wide	SELECT STRUCTURAL No. 1 No. 2	These grades are designed to fit those engineering applications where higher bending strength ratios are needed in light framing sizes. Typical uses would be for trusses, concrete pier wall forms, etc.
STRUCTURAL JOISTS & PLANKS 2 to 4" Thick 5" and Wider	SELECT STRUCTURAL No. 1 No. 2 No. 3	These grades are designed especially to fit in engineering applications for lumber five inches and wider, such as joists, rafters and general framing uses.

the same five elements are used as for dimension lumber.

Boards are separated into *Select* and *Finish* grades for a wide range of high-appearance applications. There are also five *Common* grades and five *Alternate Board* grades.

The Select grades are *B & Better*, *C Select* and *D Select*. For Idaho White Pine these grades are labeled *Supreme*, *Choice* and *Quality*. The Finish grades are *Superior*, *Prime* and *E Finish*.

For Common grades, the usual practice is to combine *1 Common* and *2 Common* and market it as *2 & Better Common*. Designations for common grades of Idaho white pine are *Colonial*, *Sterling*, *Standard Utility* and *Industrial* instead of 1 through 5 Common.

The Alternate grades are *Select*, *Merchantable*, *Construction*, *Standard*, *Utility* and *Economy*.

Size, grade, and species must all be considered in calculating allowable spans and spacing for beams, joists, and decking. These factors must also be considered in determining post size. Using the charts on pages 132-133, you can see that stronger species such as Douglas fir, Southern pine and Western larch may span wider distances than other species. These species are recommended for beams and joists, provided they are properly treated with a wood preservative. All species can be used for deck boards. Use pressure-treated wood or a naturally resistant species for posts.

Plywood Specifications—There are several rules to follow if you use plywood for a deck surface:

• Use only exterior-grade plywood.
• Choose a thickness that will span joists without being springy.
• Support plywood edges with joists or blocking.
• Cover exposed plywood edges with fascia boards or molding strips.
• Make sure the plywood surface has a sufficient slope for drainage.

The following is a description of exterior plywood grades and uses. If the plywood deck surface will be covered, appearance is not that important. If the plywood is to be painted, the surface should be smooth. If the plywood is to be clear-finished or stained, use a premier, or fine-appearance grade.

A-A Exterior—Premier quality. Use where fine appearance of both sides is desirable. Sanded smooth, easy to paint or finish.

A-B Exterior—One side premier quality. Back side nice but may have some tight knots. Good side can be stained or finished.

A-C Exterior—One side premier quality. Back side has tight knots, knotholes and minor defects.

B-B Exterior—Both sides reasonably smooth. Use where fine appearance is not essential. Easy to paint or finish.

B-C Exterior—One side reasonably smooth. Some splits and knotholes on the back. Can be painted if rough spots are filled with wood filler and sanded.

C-C Exterior—Rough and difficult to paint. Use only if surface will be covered.

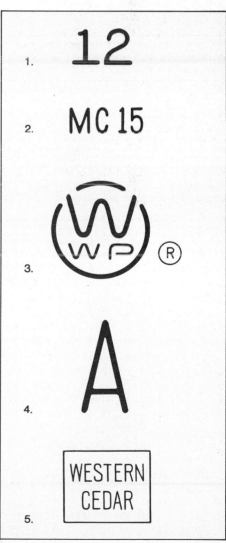

These are the marks you can expect to find on a typical lumber grade stamp. They are: 1. Number of the manufacturer. 2. Moisture content. 3. Organization which certified grading, in this case Western Wood Products Association. 4. Grade. 5. Species of wood.

Deck Spans And Spacings

The charts on these pages give the correct spans and spacings for posts, beams, joists, deck boards and plywood of various sizes. Use the charts to design your deck and to choose the proper wood species for the deck's span and spacing requirements.

[1]Based on 40 pounds per square foot deck live load plus 10 pounds per square foot dead load. Grade is Standard and Better for 4x4-inch posts and No. 1 and Better for larger sizes.

[2]Group 1—Douglas fir-larch and Southern pine; Group 2—Hem-fir and Douglas-fir south; Group 3—Western pines and cedars, redwood, and spruces.

[3]Example: If the beam supports are spaced on 8'-6" centers and the posts are on 11'-6" centers, then the load area is 98 square feet. Use next larger area 108.

MINIMUM POST SIZES (wood beam supports) [1]

Species group [2]	Post Size (in)	Load area [3] beam spacing x post spacing (sq. ft.)									
		36	48	60	72	84	96	108	120	132	144
1	4x4	Up to 12 ft. heights		Up to 10 ft. heights	Up to 8 ft. heights						
	4x6					Up to 12 ft. heights		Up to 10 ft.			
	6x6									Up to 12 ft.	
2	4x4	Up to 12 ft.	Up to 10 ft. heights	Up to 8 ft. heights							
	4x6			Up to 12 ft. heights	Up to 10 ft. heights						
	6x6						Up to 12 ft. heights				
3	4x4	Up to 12 ft.	Up to 10 ft.	Up to 8 ft. heights	Up to 6 ft. heights						
	4x6		Up to 12 ft.	Up to 10 ft.	Up to 8 ft. heights						
	6x6				Up to 12 ft. heights						

MINIMUM BEAM SIZES AND SPANS [1]
Beams are on edge.
Spans are center-to-center distances between posts or supports.

Species group [2]	Beam size (in)	SPACING BETWEEN BEAMS [3] (ft.)								
		4	5	6	7	8	9	10	11	12
1	4x6	Up to 6'								
	3x8	Up to 8'								
	4x8	Up to 10'	Up to 9'	Up to 8'	Up to 7'	Up to 6'				
	3x10	Up to 11'	Up to 10'	Up to 9'	Up to 8'	Up to 7'	Up to 6'			
	4x10	Up to 12'	Up to 11'	Up to 10'	Up to 9'	Up to 8'	Up to 7'			
	3x12		Up to 12'	Up to 11'	Up to 10'	Up to 9'		Up to 8'		
	4x12					Up to 11'	Up to 10'	Up to 9'		
	6x10						Up to 11'	Up to 10'		
	6x12						Up to 12'			
2	4x6	Up to 6'								
	3x8	Up to 7'		Up to 6'						
	4x8	Up to 9'	Up to 8'	Up to 6'						
	3x10	Up to 10'	Up to 9'	Up to 8'	Up to 7'		Up to 6'			
	4x10	Up to 11'	Up to 10'	Up to 9'	Up to 8'		Up to 7'		Up to 6'	
	3x12	Up to 12'	Up to 11'	Up to 10'	Up to 9'		Up to 8'	Up to 7'		
	4x12					Up to 10'		Up to 9'		
	6x10						Up to 9'	Up to 8'		
	6x12					Up to 12'	Up to 11'	Up to 10'		
3	4x6	Up to 6'								
	3x8	Up to 7'	Up to 6'							
	4x8	Up to 8'	Up to 7'	Up to 6'						
	3x10	Up to 9'	Up to 8'	Up to 7'	Up to 6'					
	4x10	Up to 10'	Up to 9'	Up to 8'		Up to 7'		Up to 6'		
	3x12	Up to 11'	Up to 10'	Up to 9'		Up to 7'		Up to 6'		
	4x12	Up to 12'	Up to 11'	Up to 10'	Up to 9'	Up to 8'			Up to 7'	
	6x10		Up to 12'	Up to 11'	Up to 10'		Up to 8'	Up to 8'		
	6x12			Up to 12'	Up to 11'	Up to 10'	Up to 9'	Up to 8'		

[1]Based on 10 pounds per square foot deck live load plus 10 pounds per square foot dead load. Grade is No. 2 or Better, No. 2, medium-grain Southern pine.
[2]Group 1—Douglas fir-larch and Southern pine; Group 2—Hem-fir and Douglas-fir south; Group 3—Western pines and cedars, redwood and spruces.
[3]Example: If the beams are 9 feet, 8 inches apart and the species is Group 2, use the 10-ft. column; 3x10 up to 6-ft. spans, 4x10 or 3x12 up to 7-ft. spans, 4x12 or 6x10 up to 9-ft. spans, 6x12 up to 11-ft. spans.

MAXIMUM ALLOWABLE SPANS FOR DECK JOISTS[1]

Species Group[2]	Joist size (inches)	JOIST SPACING (INCHES)		
		16	24	32
1	2x6	9'-9"	7'-11"	6'-2"
	2x8	12'-10"	10'-6"	8'-1"
	2x10	16'-5"	13'-4"	10'-4"
2	2x6	8'-7"	7'-0"	5'-8"
	2x8	11'-4"	9'-3"	7'-6"
	2x10	14'-6"	11'-10"	9'-6"
3	2x6	7'-9"	6'-2"	5'-0"
	2x8	10'-2"	8'-1"	6'-8"
	2x10	13'-0"	10'-4"	8'-6"

[1]Joists are on edge. Spans are center to center distances between beams or supports. Based on 40 pounds per square foot deck live loads plus 10 pounds per square foot dead load. Grade is No. 2 or Better; No. 2 medium grain Southern pine.
[2] Group 1—Douglas-fir-larch and Southern pine; Group 2—Hem-fir and Douglas-fir south; Group 3—Western pines and cedars, redwood and spruces.

MAXIMUM ALLOWABLE SPANS FOR SPACED DECK BOARDS[1]
Maximum allowable span (inches)[3]

Species group[2]	LAID FLAT				LAID ON EDGE	
	1x4	2x2	2x3	2x4	2x3	2x4
1	16"	60"	60"	60"	90"	144"
2	14"	48"	48"	49"	78"	120"
3	12"	42"	42"	42"	66"	108"

[1]These spans are based on the assumption that more than one deck board carries normal loads. If concentrated loads are a rule, spans should be reduced accordingly.
[2]Group 1—Douglas fir-larch and Southern pine; Group 2—Hem-fir and Douglas-fir south; Group 3—Western pines and cedars, redwood and spruces.
[3]Based on Construction grade or Better (Select Structural, Appearance, No. 1 or No. 2).

PLYWOOD SPECIFICATIONS
Recommended grades, minimum thicknesses, and nailing details for various spans and species groups of plywood decking[1]

Plywood species group[2]	Panel thickness in inches[3, 4] For maximum spacings between supports (inches)			
	16	20	24	32 or 48
1	1/2	5/8	3/4	1-1/8
2 & 3	5/8	3/4	7/8	1-1/8
4	3/4	7/8	1	5

[1]Recommended thicknesses are based on Underlayment Exterior (C-C Plugged) grade. Higher grades, such as A-C or B-C Exterior, may be used. 19/32" plywood may be substituted for 5/8-inch and 23/32" for 3/4-inch.
[2]Plywood species groups are approximately the same but not identical to those shown for lumber in the other tables on these pages. Therefore, in selecting plywood, one should be guided by the group number stamped on the panel.
[3]Edges of panels shall be tongue-and-groove or supported by blocking.
[4]Nailing details: Size—6d annular ring or screw nails, except 8d for 7/8" or 1-1/8" plywood on spans 24 to 48". Spacing—6" along panel edges. 10" along intermediate supports (6" for supports 48" on center). Rust-resistant nails are recommended where nail heads are to be exposed. Nails should be set 1/16" (1/8" for 1-1/8" plywood).
[5]Not permitted.

HOW TO ORDER AND BUY LUMBER

Lumber is sold in several ways. Some stores sell lumber by the *linear* or *running foot*, others by the *board foot*. A linear foot is a foot-long piece of wood in whatever dimension you want. It is priced accordingly. A board foot is equal to the amount of wood in a piece 1 foot wide by 1 foot long by 1 inch thick. An 18-inch long 2x4 or a 1-foot long 2x6 equals 1 board foot.

Lumberyards or home-improvement centers that sell by the board foot usually have a price per 1000 board feet for each size, species and grade of lumber. This price is divided by the number of board feet you order.

Here's how you determine the board footage of your deck material. For each size, multiply the width in inches by length in inches by thickness in inches. Then multiply this figure by the number of pieces you're using. Divide the total by 12. Remember, figures used in this formula are nominal, not actual sizes. Use these figures when ordering lumber, not to determine the actual board footage in your deck for sizing purposes.

Estimating Amounts—Compile an estimate sheet with columns for the sizes, lengths, and quantities of the materials you need. In areas where you expect lumber waste or loss, such as the deck surface, add 5 to 10% to the estimated amount.

Refer to your plans to estimate lumber sizes and amounts. If you haven't already determined how much lumber you'll need, use the formula below. In the formula, *deck width* is the measurement perpendicular to the direction the boards will run.

Formula:

For 2x4s—multiply the deck width in feet by 3.3 and round to the next-highest foot.

For 2x6s—multiply the deck width in feet by 2.1 and round to the next-highest foot.

If you are going to use 2x4s or 2x6s on edge, substitute 7.1 for 3.3 or 2.1 in the equation.

When you order your material, make the best use of standard lumber sizes. For example, if you're covering a 12-foot span, you may find 8 footers less expensive than 12 footers. Buy 8 footers, cut some of them into 4-foot lengths and use them in combination to make 12-foot lengths. Make sure joists are positioned under joints. Stagger joints on alternate runs.

Here's another example. Your deck has an 11-foot span. Don't waste a foot each time by cutting off 12-footers. Instead, buy 8-footers and 6-footers and cut the 6-footers into 3-foot lengths. Use 8-footers and 3-footers in combination, staggering joints.

The type of store will make a difference in your lumber shopping. An ordinary, old-fashioned lumberyard generally has the widest selection of lumber grades, sizes and species. They usually have knowledgable personnel to advise you. Home-improvement centers and self-service building emporiums supply a plethora of building and household items but their lumber stock is usually limited. However, these outlets often run sales. If you're lucky, you may be able to pick up part of your decking lumber at an appreciable discount.

In all cases, do some comparative shopping before you buy. Check not only prices, but the quality of the lumber. You may find it economical to buy lumber from several different outlets. On the other hand, some lumberyards may give you a volume discount if you buy all your materials from them.

Many lumberyards have experienced personnel who will help you estimate and select materials. Some yards give discounts or "contractors" prices if you buy all your materials from them.

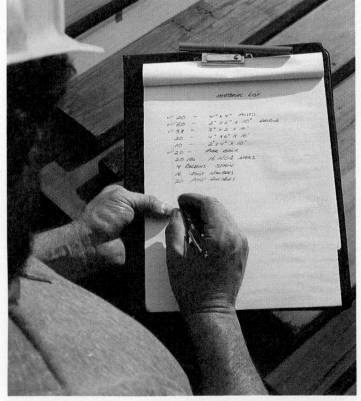

Make a complete list of needed materials before you visit lumberyards and home-improvement centers. Use the list to compare prices and to make sure you don't forget anything when you go to buy.

FASTENERS

Fasteners used for deck construction include nails, screws, bolts and different metal connectors. All metal fasteners should be rust-resistant and designed for exterior use.

Nails—Here are some types of nails you may need for deck construction. *Common nails* and *box nails* are used for fastening substructure members and deck boards. *Finishing nails* and *casing nails* are used where exposed nail heads would be unsightly, such as on moldings and trim strips. *Annular ring* or *spiral* nails are sometimes used for fastening deck boards, but are more often used on plywood decking or roof coverings. These nails hold much better than box or common nails. *Duplex* nails have double heads so they can be easily removed from the wood. They're used for temporary nailing. *Concrete* nails are for nailing wood to concrete.

Use a nail three times as long as the thickness of the top board you are nailing. Nail lengths are designated by penny sizes. For example, a 6d—or six-penny—nail is 2 inches long.

Nails are generally sold by the pound. Price per pound depends on the nail type and size. The chart on page 136 tells you how many nails there are per pound of each size and type.

Nails for outdoor use should be rust-resistant. Galvanized nails are readily available and won't rust unless the galvanized coating is chipped off. You can use these for all parts of the deck.

Aluminum nails resist rust better than galvanized nails, but they're more expensive and bend easily. You might consider using aluminum nails for the deck surface if it is of redwood or a soft pine. Stainless-steel nails best resist rust, but are extremely expensive and not readily available in most retail outlets.

Nails with sharp points will split wood. It's best to blunt the nail points with a hammer before driving them.

Screws—When using wood screws, make sure the threaded portion of the screw is completely embedded in the bottom piece of wood. Predrill a clearance hole in the top piece of wood for the threadless part of the screw, called the *shank*. Drill a pilot hole for the threaded part. Special drill attachments, called *screw-pilot bits,* predrill both holes and a countersink hole for flat-head wood screws.

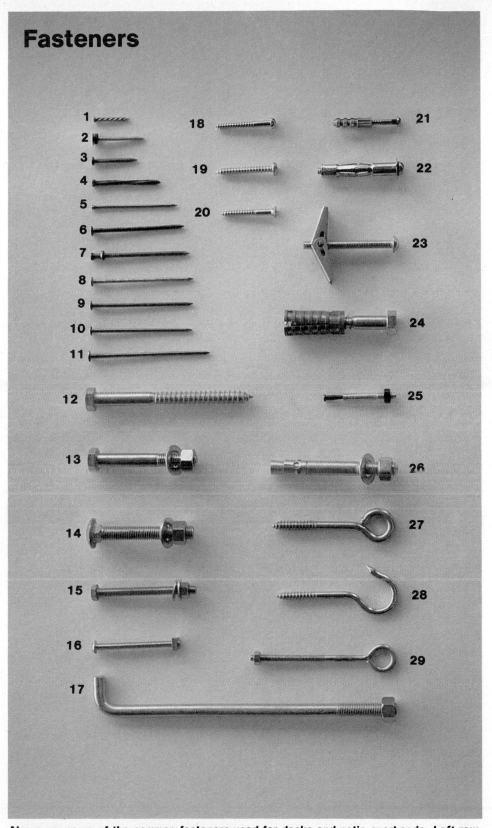

Fasteners

Above are some of the common fasteners used for decks and patio overheads. *Left row, top to bottom:* 1. Screw nail. 2. Aluminum twist nail with neoprene washer (for corrugated fiberglass panels). 3. Joist-hanger nail (used with metal connectors). 4. Masonry nail. 5. 10d finish nail. 6. 16d vinyl-coated sinker. 7. Duplex nail. 8. 16d galvanized box nail. 9. 16d common nail. 10. 16d box nail. 11. 20d spike. 12. Lag screw. 13. Machine bolt. 14. Carriage bolt. 15. Full-thread machine bolt. 16. Stove bolt. 17. J-bolt. *Center row, top to bottom:* 18. Round-head wood screw. 19. Oval-head wood screw. 20. Flat-head wood screw. *Right row, top to bottom:* 21. Plastic anchor with screw. 22. Molly bolt. 23. Toggle bolt. 24. Lead expansion shield with lag screw. 25. Red-head stud anchor. 26. Stud anchor. 27. Eye screw. 28. Screw hook. 29. Eye bolt.

NAIL SIZES AND AMOUNTS

Penny Size	Length	Number Per Pound
2d	1"	
Common		876
Box		1010
Finish		1351
3d	1-1/4"	
Common		568
Box		635
Finish		807
4d	1-1/2	
Common		316
Box		473
Finish		548
5d	1-3/4	
Common		271
Box		406
Finish		500
6d	2"	
Common		181
Box		236
Finish		309
7d	2-1/4"	
Common		161
Box		210
Finish		238
8d	2-1/2"	
Common		106
Box		145
Finish		189
9d	2-3/4"	
Common		96
Box		132
Finish		172
10d	3"	
Common		69
Box		121
Finish		132
12d	3-1/4"	
Common		64
Box		94
Finish		113
16d	3-1/2"	
Common		49
Box		71
Finish		90

Bolts—Bolts should be as long as the total of both pieces being joined, plus 1 inch. Predrill a bolt hole in the wood 1/16-inch larger than the bolt.

Metal Connectors—It is possible to build a sound deck using only nails, bolts and screws. The preferred way to build a deck is to use them in conjunction with preformed *metal connectors*. Metal connectors include metal post anchors, post caps, joist hangers, beam saddles and other assorted straps and hangers. As you read through this chapter you'll see how these connecters are used to connect different deck members. The connectors will be identified by their proper names in the illustrations.

Metal connectors not only add rigidity to the deck structure, they help prevent lumber from twisting or warping after the deck is built.

EXTERIOR FINISHES

For exposed flat surfaces of a deck, a *natural finish* containing a water-repellent preservative is preferable to paint. Natural finishes include semitransparent and solid-color stains, and clear penetrating finishes. These penetrate wood but do not form a surface film as paint does. Surface films tend to rupture after a period of time, allowing decay-causing organisms to enter the wood.

Clear finishes allow the natural color and grain of wood to show through. They are generally used to repel moisture and delay weathering.

Semitransparent stains allow the natural grain and texture of the wood to show through. They don't last as long as solid-color stains, but they're easy to reapply because the wood requires no surface preparation.

Solid-color stains are more durable, especially on rough or weathered wood, such as beams and posts. These stains hide grain patterns completely but allow wood texture to show through.

Paints completely hide wood color, grain and texture. You have a choice between exterior latex and exterior oil-base, or alkyd paint. Both paints come in flat and glossy finishes. Latex paints do not stick to wood as well as oil-base paints, and are best used on new, relatively smooth wood. Oil-base paint is preferable on rough or previously painted wood. Latex paints are generally easier to apply than oil base paints and do not require thinner or other solvents. Hands and brushes can be cleaned with ordinary water.

Special Coatings—Tough, slip-resistant, watertight coatings with *neoprene*, *silicone* and *rubber-based finishes* can be used on plywood floors. However, they are difficult to work with and should be applied by skilled workers. If outdoor carpeting is used on plywood surfaces, the plywood should be pressure-treated and the deck substructure well-ventilated.

A clear, water-repellent finish will help extend the life of deck boards. It will also restore some of the natural color to a weathered deck surface. Use a penetrating finish, not one that forms a surface film.

Anatomy Of A Deck

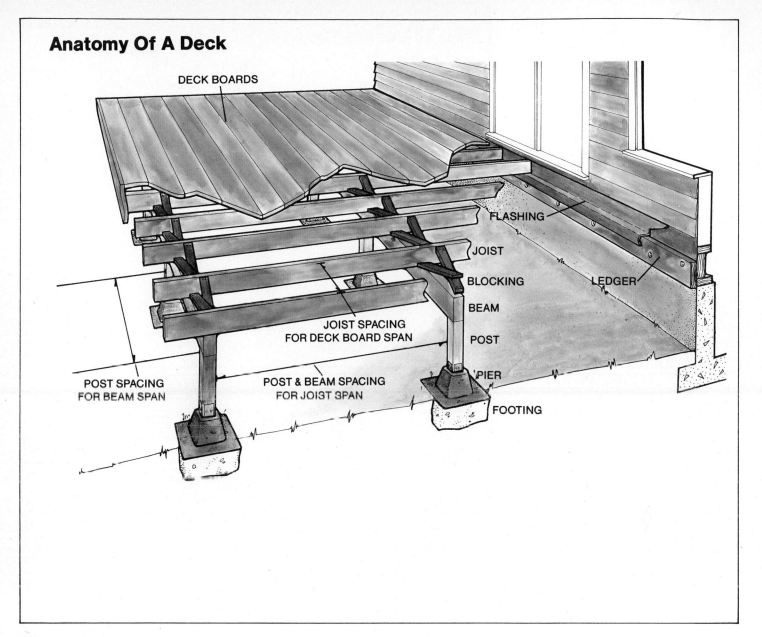

DECK BOARDS

FLASHING

JOIST

BLOCKING

LEDGER

BEAM

JOIST SPACING
FOR DECK BOARD SPAN

POST

POST SPACING
FOR BEAM SPAN

POST & BEAM SPACING
FOR JOIST SPAN

PIER

FOOTING

Deck Building Sequence

The following is a step-by-step guide to building a deck. It includes building variations in deck substructures and surfaces. You'll also find building details for railings and benches.

STEP 1—SITE PREPARATION

Before you start construction, make sure the site has adequate drainage. You may have to do some grading to provide a slight slope for water runoff.

If roof gutters drain into the area, reroute the water through an underground drain line. Dig an 18-inch-deep trench from the gutter outlet to a dry well or drainage field. Lay 4-inch drain tile in the trench. Butt tiles loosely together and cover joints with a strip of 30-pound asphalt roofing felt. Then cover the trench with about 8 inches of drain rock and a layer of dirt.

Be sure to check local regulations on the use of dry wells. Dig the well 2 to 3 feet in diameter. It should have straight sides and be deep enough to reach a good drainage area. Fill the well with 2- to 4-inch-diameter stones.

Remove any weeds, dead roots, wood scraps, or other organic matter from the deck site. These all contribute to decay. Wood scraps also attract termites. Do not treat the soil with weed killers or insecticides until you are ready to lay the deck boards. See page 150.

STEP 2—ATTACH LEDGER

If the deck will be attached to the house or other building, you will need to attach a *ledger* to the house wall. The ledger, in turn, is attached to the deck joists.

The ledger is attached through the house wall directly to interior floor framing or to the masonry foundation, as shown in the drawings on page 139. The ledger must be attached properly, as it is the starting point for the deck substructure.

For small decks, ledgers can be of the same size lumber as the deck joists. For large or elevated decks, use a 2x8 or 2x10, depending on code requirements.

The length of the ledger will depend on how you will attach the joists to it. If the end joists of the deck will be attached to the ledger with joist hangers, allow several inches of nailing space at each end of the ledger for attaching the outer sides of the hangers. If the end joists will be attached to the ends of the ledger, cut the ledger shorter to compensate for the width of the joists.

Position the ledger so the finished deck surface will be an inch or more below existing or proposed door openings. Measure down from the sill one inch, plus the thickness of the deck boards. Add the width of the joist if it will rest on top of the ledger. This positioning keeps water from entering the house through the doorway.

On wood-frame houses, use lag screws to attach the ledger directly to the floor header. See drawing on page 139, top left. Do not attach the ledger to studs or other house-framing members. Use this method for attaching ledgers to headers of second-story floors. The drawing on page 139, top right, shows how to attach a ledger to a wood-frame house with stucco walls.

If you're attaching the ledger to a solid masonry wall or foundation, use one of the methods shown in the bottom drawings on page 139.

On some houses with basements, the floor header is exposed. If the floor header in your house is not, you'll have to locate its position on the outside wall. To do this, you must first locate the inside floor level. This level can be easily located using a nearby doorway. If no doorway is nearby, measure from a window sill to the floor. Then transfer these measurements to the outside wall. Floor headers are usually 2x10s with subflooring and flooring on top of them. Measure 6 inches down from the floor level line marked on the outside wall. This height should be the centerline of the header, the target for your lag screws.

To attach the ledger, first brace it against the wall. On wood walls, tack the ledger to the wall with one nail at the center. Use a carpenter's level to level the ledger, then tack nails in both ends. Use duplex nails so you can easily remove them when the ledger is bolted in place.

For stucco or masonry walls, use 2x4s with cleats set against stakes for angled supports. Predrill holes in the ledger for lag screws or expansion bolts. Use a masonry bit to drill holes through stucco into the header. For solid masonry walls or foundations, use a masonry bit to drill holes for expansion anchors or stud anchors. See drawings on page 139.

How To Locate The Ledger

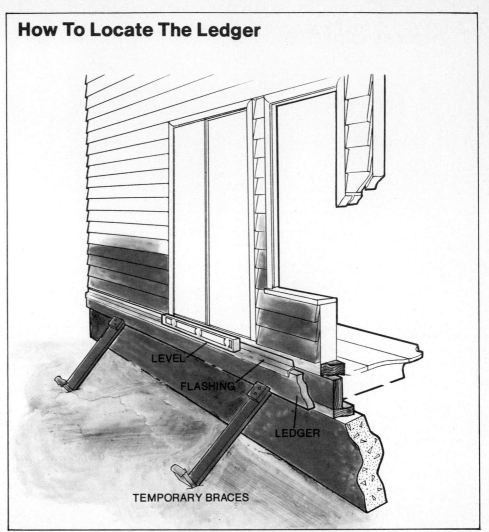

LEVEL

FLASHING

LEDGER

TEMPORARY BRACES

Locate the ledger so the finished deck surface will be 1" below the door sill, as described in the text. If there is no door, measure from an inside window sill to the floor. Transfer this measurement outside and add 6 inches. This should be the center line of the floor header. On wood-frame houses, position the ledger so it can be attached to the floor header with lag screws. Be sure to maintain the required distance below the door opening when positioning the ledger. Once the ledger is in position, brace and level it as shown in the drawing. Attach it to the house wall with the appropriate fasteners, as shown in the drawings on page 139.

Attach screws or bolts, remove bracing and recheck level. If the ledger is slightly off level, loosen the bolts or screws slightly and adjust the ledger. Retighten bolts or screws. If this doesn't solve the problem, take the ledger off and start over. Plug and seal all holes before replacing the ledger.

To help prevent decay, ledgers should be flashed with aluminum or galvanized metal flashing, as shown in the drawing above. Flashing prevents moisture from being trapped between the ledger and house wall.

To make galvanized metal or aluminum flashing, add joist thickness, decking thickness, and ledger width plus 1 inch. The total will be the width of your flashing. Starting one inch below the top of the ledger, crimp the flashing so it runs over the top of the ledger and up the wall flush with the top of the proposed deck surface. To make 90° bends in flashing, sandwich it between two 2x4s and clamp together. Use a hammer and short block of wood to form the crimp. If the house has horizontal lap siding or shingles, cut the flashing wide enough to be inserted under the house siding just below the deck surface level. The flashing is then nailed to the wall studs. On flush walls, the top can be sealed with a bead of silicone or butyl-rubber caulk. On a stucco house, snap a chalk line just below the deck surface level. Use a portable circular saw with a carborundum blade to saw a 3/8-inch-deep kerf into the stucco. Bend the flashing to fit into the kerf. Seal with silicone or butyl-rubber caulk.

Four Ways To Fasten A Ledger

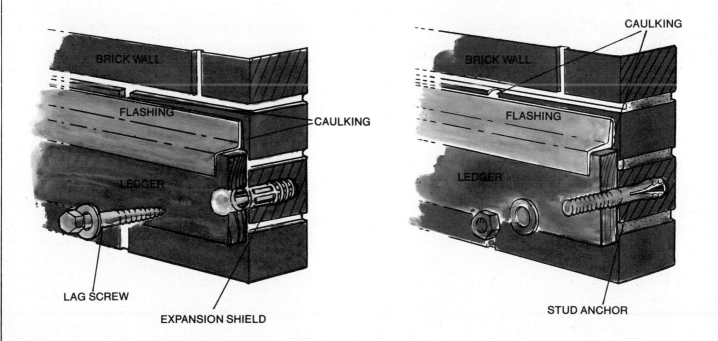

Use lag screws to fasten ledger to wood-frame houses. Drill clearance holes in the ledger for the lag-screw shank. Brace the ledger against the wall. Drill a smaller pilot hole through the wall into the floor header for the lag screw threads. Use a wrench to tighten lag screws.

For wood-frame houses with stucco walls, attach the ledger with lag screws as shown. Use a masonry bit to drill clearance holes through the stucco. Use a wood bit to drill pilot holes in the floor header behind the stucco. Use lag screws with shanks long enough to go through both the ledger and stucco surface. Flashing is fit into a groove, or *kerf,* cut into the stucco, then sealed with butyl caulk or silicone seal.

One way to attach a ledger to a masonry wall is with lag screws and lead expansion shields. Position the ledger and drill holes through the ledger into the masonry wall. Insert the expansion shields as shown then fasten the ledger with lag screws.

Another way to attach a ledger to a masonry wall is with stud anchors. Position the ledger and drill holes through it. Mark hole locations on wall, remove ledger and fasten stud anchors to wall as shown. Replace ledger and tighten nuts.

STEP 3—LOCATE FOOTINGS AND PIERS

A deck foundation consists of concrete piers or metal post anchors set in concrete footings. Local code requirements will vary on footing size and depth. Check codes before digging any holes. Generally, a footing should be twice the width of the pier it supports. Weak footings can ruin the whole deck.

Refer to your working drawings to locate footing positions. The posts will be centered directly over the footings. Refer to the chart on page 132 to make sure you have the correct post spacing to support your deck.

The first step in locating footings is to locate the outside corners of your deck. To do this, construct *batter boards* several feet outside deck corner locations as shown in the drawing below. If the deck is freestanding, you'll need 4 sets of batter boards on which to attach leveling strings—one set at each corner. If you're using a ledger, you'll only need 2 sets of batter boards, as shown in the drawing.

The stakes for the batter boards must be firmly embedded in the ground. The assembly must hold the pressure of taut strings. Level the crosspieces on the batter-board assemblies. Then level the batter board tops to each other. To do this, attach string between the batter boards, as shown in the drawing below. Use a line level to level each string. Adjust the batter-board heights until all the strings are level.

Here's how to square the corners of a freestanding deck. In this method, the strings outline the perimeter of the deck, or the outside faces of end joists and joist headers. The footings should be located accordingly.

Starting at one corner point where two strings intersect. Measure 3 feet down one string, 4 feet down the other. Mark these locations with a felt-tip pen or a short piece of string.

Use a tape measure or folding rule to measure diagonally between the two marks. Have a helper adjust the strings on the batter board until the diagonal measurement equals 5 feet. The intersecting strings should now form an exact 90° angle. Leave these two strings in place and adjust the other two strings so all four corners are square. Once all four corners are established, measure diagonally across the square in both directions. If both diagonal measurements are the same, the strings form a perfect square.

How To Locate Footings And Piers

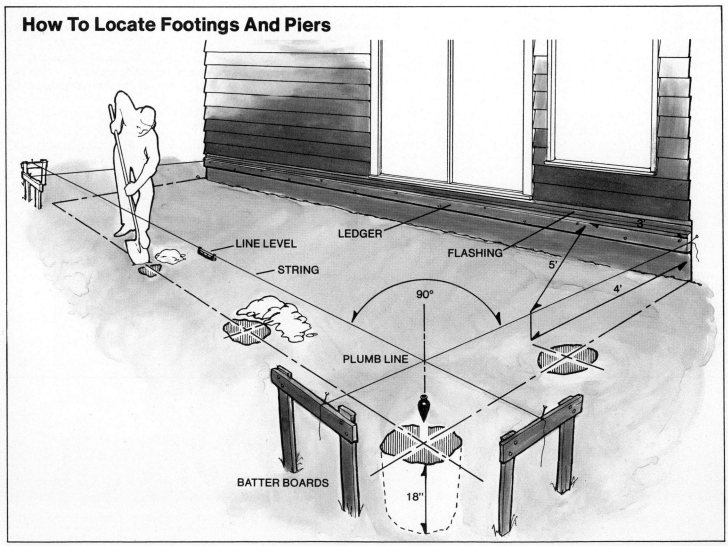

Outline the deck with string. Use batter boards to keep string taut and level. Make sure deck outline is square with the 3-4-5 triangle method, as shown. Use the charts on page 132 to determine how far to space footings. Dig holes at least 18 inches deep in appropriate positions.

Here's how to square deck corners if your deck has a ledger. At one end of the ledger, mark the location where the outside face of the end joist will butt against the ledger or house wall. If the joist will be attached to the ledger end, the outside face of the joist will be 1-1/2 inches beyond it. Attach string to the ledger end at the mark you just made. Pull the string taut and attach the other end to the batter board.

Mark the string 4 feet out from the ledger end with a felt-tip pen or short piece of string. Mark the ledger 3 feet from the end. If the diagonal measurement between the two marks equals 5 feet, the string should be squared to the ledger. Double-check squareness with a framing square. Repeat this process at the other end of the ledger. Finally, run a third string between the two batter boards and square it to the first two strings.

Check overall squareness by taking diagonal measurements from corner to corner. Both measurements should be equal.

Once the strings are squared, use them to locate footings around the deck perimeter. Keep in mind that the strings mark the perimeter of the finished deck. Refer to your plans to determine footing positions within string boundaries.

Some deck designs call for *cantilevering* the structure several feet out from the supporting posts. This construction method is often used on hillside decks, and on low-level decks to hide unsightly footings and piers. It gives the deck the appearance of floating over the ground. Joists and beams can be cantilevered a distance equal to 1/2 their allowable spans between supporting deck members. For example, a joist with an 8-foot span can be cantilevered 4 feet out from the end beam. Check local codes. If you're cantilevering the deck, locate footings accordingly.

If the outside edges of posts, beams and joists will be flush to each other, the points where the strings intersect form the outside edges of the corner posts. Piers and footings should be placed accordingly. To locate corner footings, use a plumb line to mark the ground directly beneath the intersecting strings, as shown in the drawing at left. Mark the spots with stakes until you're ready to dig footing holes.

To locate perimeter footings between the corner footings, measure down the strings and use a plumb line to mark footing positions on the ground. Footings within the deck perimeter can be located by centering strings between perimeter footing locations and using the squaring and measuring techniques just described.

STEP 4—INSTALL FOOTINGS AND PIERS

By simple definition, footings are holes filled with concrete. On some footing systems, the footings are poured 3 to 6 inches below ground level and concrete piers, either precast or poured-in-place, are embedded in the footings. On other footing systems, posts are connected directly to the footings with post anchors or other fasteners. On raised decks where exposed footings would be unsightly, the footings may be sunk below grade level and the posts embedded in them or attached to them. A thin layer of earth covers the footing and post. See drawings on page 143.

Before you dig holes for the footings, find out what size they must be. Footing and pier size depends on deck weight and soil conditions. A footing hole 12 inches square and 8 inches deep is adequate for most decks built on stable soil. The footings themselves need only be 6 inches thick. Under ideal conditions, footings may not be required at all. Precast concrete piers can be set in a shallow hole with a few shovels full of wet concrete placed around them.

In cold climates, codes often require footings to be sunk below frost lines. This eliminates deck heaving due to freezing and thawing soil. If the deck site is on loose soil or fill, you'll have to dig down to undisturbed soil and put the footing in it. If loose soil is too deep, rent a power compactor to compact the dirt before pouring the footing.

Aboveground footings and piers should provide anchorage so posts resist lateral movement and uplift which can occur during high winds. This is done by attaching or embedding post anchors, anchor bolts, drift pins or other fasteners to the tops of footings or piers. See "Attach Posts To Footings Or Piers," page 143. Aboveground footings and piers should extend at least 6 inches above ground level to prevent post decay.

Once you know the necessary size and depth of the footings, select a footing system that best suits your deck's design and structural requirments. Several common footing systems are described here.

Footings And Precast Piers—Footings are poured to within 3 to 6 inches of ground level. While the concrete is still wet, precast piers are embedded 1 to 2 inches into the footing. You can buy precast piers or make your own—the cost is about the same. Most precast piers come with nailing blocks. Some have drift pins or post anchors embedded in them, but these piers aren't always readily available. You can attach post anchors to pier nailing blocks or embed them in piers you make yourself.

You can make wooden forms for precast piers, or use large metal or plastic containers. Wood for forms should be at least 1/2-inch thick. If you use containers for forms, cut them lengthwise, then tie them back together with wire or twine before pouring concrete into them. This allows forms to be easily removed once the concrete has set. Precoat forms with motor oil to keep concrete from sticking to them.

Another type of precast pier is a concrete column. These are used when footings are unusually deep and it would be impractical to pour the footing up to ground level. Columns can double as posts for low-level decks. If the column will extend 3 feet or more above ground, it should be reinforced with metal rods.

You can make concrete columns using fiber tubes manufactured for that purpose. You can buy fiber tubes at masonry supplies. See drawing on page 142.

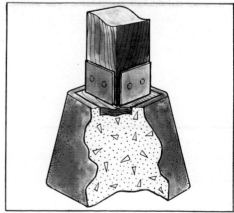

Precast piers have nailing blocks to attach a metal post anchor as shown here. For low-level decks, beams can be toenailed to pier nailing blocks, or fastened with metal connector.

Footings And Formed Piers—In this system, footings and piers are poured successively to form one solid unit. First, the footing is poured. While the concrete is still wet, two or three 8 to 10-inch lengths of metal reinforcing rod are embedded in the footing so about 4 inches of each rod extend above it.

A wooden form is then set into the still-wet concrete, leveled and filled with concrete. A post anchor, drift pin or other fastener is embedded in the formed pier while the concrete is still wet. When the concrete sets, the form is removed.

You can make wooden forms for piers out of 1-inch-thick board scraps, or 1/2-inch or thicker plywood. Simply build a square frame with inside dimensions that match the outside dimensions of the pier. Before pouring the piers, coat the form with clean motor oil so the concrete won't stick to the form.

Hollow masonry blocks can be used as permanent forms for piers. The blocks are placed on the footings while the concrete is still wet. The core holes are then filled with concrete and the block is leveled. Reinforcing rods are inserted in the holes to link the pier to the footing. A nailing block is then attached to the top of the block with lag screws or J-bolts while the concrete is still wet. See drawing below.

Below-Grade Footings—You may use two types of below-grade footings for treated posts or poles. The first is a precast footing or pier placed at the bottom of the footing hole. It usually is sunk 2 to 3 feet to resist lateral movement. In cold climates, it may be sunk 4 or more feet to reach below frost level. The post is attached to it by a drift pin embedded in the precast footing. The minimum footing size in stable soil is 12x12x8''.

The other type of below-grade footing is poured in place. This provides sturdier post support than the method above. Posts are aligned, plumbed and suspended 6 to 8 inches above the bottom of footing holes. Concrete is then poured into the holes. The concrete should cover the bottom 8 to 12 inches of the post and extend 6 to 8 inches below it. The remainder of the hole is packed with gravel or tamped earth. Use only pressure-treated posts or poles with below-ground footings. The pressure-treated lumber must be designated for below-ground use. See page 127.

Mixing And Pouring Concrete—The standard concrete mixture for footings and piers consists of 1 part Portland cement, 2 parts clean sand and 3 parts washed gravel. If you have only a few footings, you can use dry, ready-mix concrete in sacks. A 90-pound sack of dry-mix equals 2/3 cubic foot of set concrete. This will make a 12x12x8'' footing. One cubic foot will make a 12x12x12'' footing or pier, 1/2 cubic foot, a 12x12x6'' footing or pier. If you have very many footings to pour you can either mix your own concrete from its component parts or have the concrete delivered in a truck, ready to pour. In some localities, you can buy ready-to-pour concrete from a dealer and haul it home in a trailer.

To estimate how much concrete you need, first determine how many cubic feet are needed for each footing, and for each pier if you're making these yourself. Multiply this amount by the number of footings and piers needed. For more complete information on estimating amounts of concrete and its component parts, see page 108.

Mix dry concrete mix, or concrete components with water in a wheelbarrow or mixing trough. Use a shovel or ordinary garden hoe for mixing. The mixture consistency should be *stiff* and *plastic*, that is, not too soupy or too dry. Your wheelbarrow should hold between 4 and 6 cubic feet of concrete. Use about 5 gallons of water for 4 cubic feet of concrete, 6 gallons

Pouring Footings And Piers

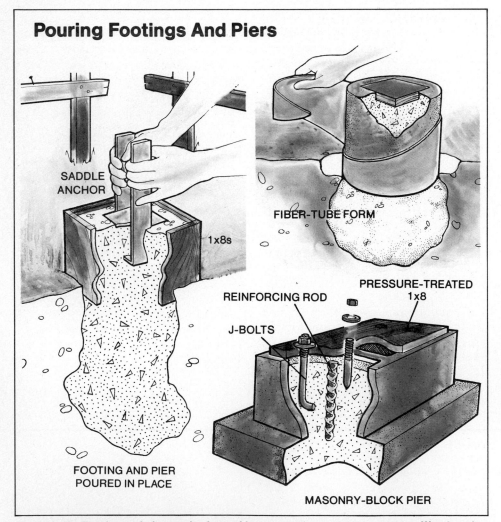

SADDLE ANCHOR

1x8s

FIBER-TUBE FORM

FOOTING AND PIER POURED IN PLACE

REINFORCING ROD

J-BOLTS

PRESSURE-TREATED 1x8

MASONRY-BLOCK PIER

Above Left: Footing and pier can be formed in one continuous pour as shown. Wooden pier forms should be at least 6 inches deep. Dig footing hole to correct depth and fill with wet concrete. Position pier form on footing and fill with concrete. Position post anchor in wet concrete. Anchor shown is called a *saddle anchor.* When concrete has set, remove wooden form. Above Right: Prefabricated cardboard tubes are used as forms for making concrete columns. These are generally used where footings must be unusually deep. The columns can double for posts on low-level decks. Below Right: A hollow masonry block can be used as a permanent pier form. For details, see text above.

for 5 cubic feet, 7-1/2 gallons for 6 cubic feet. Use a bucket for adding proper amounts of water.

Place the dry concrete components in your wheelbarrow or trough and mix thoroughly. Make a depression in the middle of the pile and add about half the total amount of water needed. Mix thoroughly. Continue mixing as you slowly add the remaining amount of water until the concrete is of proper consistency.

When the concrete is mixed, pour it into the footings. Place or pour piers on footings while the concrete is still wet. Align piers to strings on the batter boards, then level the piers and adjust them for height. If you're embedding post connectors in footings or piers, set and align the connectors before the concrete sets.

STEP 5—ATTACH POSTS TO FOOTINGS OR PIERS

Cut posts 6 to 8 inches longer than their finished height above the piers. This will allow for variations in pier height when you level and cut the post tops.

There are several connectors used to attach posts to concrete piers and footings. The more common ones are shown in the drawings at right. The connector you use will depend on the deck design and the type of piers and footings you're using. The fasteners you use should securely anchor the post to the pier or footing. This is especially important for decks in windy locations.

The most common and perhaps weakest method of post anchorage is to toenail the posts to the pier nailing blocks. For a few dollars more, you can buy metal *post anchors*, fasten them to the nailing blocks and attach the posts to them. This makes a much stronger installation. A *drift pin* embedded in the pier is another good way to fasten posts. See drawings at right and on page 141.

In wet climates, choose a fastener that keeps moisture from being trapped between post and the pier or footing. A *step flange anchor* or a *saddle anchor* keeps the post suspended above the pier or footing. This permits air circulation to keep the post end dry. A *corner angle* can also keep the post from resting directly on the pier or footing. These anchors are shown in the drawings at right.

If your deck design calls for double posts, use the double-post anchor

Connecting Posts To Piers

DRIFT PIN EMBEDDED IN PIER

STEP FLANGE ANCHOR

DOUBLE POST ANCHOR

CORNER ANGLE

BELOW-GRADE POST AND FOOTING

SADDLE ANCHOR

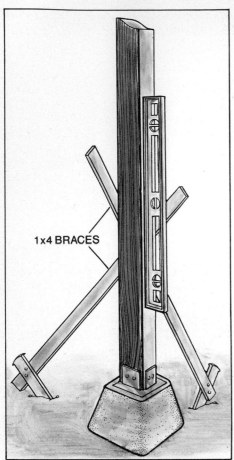

1x4 BRACES

To level post vertically, check adjacent post faces with a carpenter's level. Attach temporary braces to posts to keep them level until beams are attached.

shown in the drawing. Double posts are usually a set of 2x4s or 2x6s attached to either side of the beam.

All fittings and hardware should be rust resistant. Use pressure-treated lumber for posts and pier nailing blocks. Cut ends of pressure-treated posts should be soaked in a preservative overnight or longer. See page 127.

Set the posts on the piers or footings. Construct braces to hold them while you plumb them vertically on adjacent faces with a carpenter's level. See drawing above. Posts 2 feet or shorter probably won't need bracing. When all posts are plumb, they can be permanently fastened to the piers or footings.

STEP 6—LEVEL POST TOPS

Once all the posts are attached to the piers, you must cut the post tops so they are level with each other. This ensures even beam contact and load distribution. Use a string and line level or a length of 2x4 and a carpenter's level to level the tops. If the deck has a ledger, use it as your starting

point, as shown in the drawing below. Otherwise, use one of the corner posts. If the deck is attached to the house, the deck should slope slightly away from it, about 1 inch slope for every 10 feet of deck run.

To level posts to the ledger, run a leveled string or 2x4 from the top of the ledger across the post faces, as shown in the drawing. If the deck will slope away from the house, move the string or 2x4 down slightly on the post farthest away from the ledger. For instance, if the farthest post is 10 feet from the ledger, drop the string or 2x4 one inch.

Mark the posts where the string or bottom edge of the 2x4 intersects them. Measure down the post from these marks the width of the beam and make a second mark. Use a try square and pencil to mark the cutoff line at the second mark.

Cut post tops with a portable circular saw. If the posts are very tall, you may find it easier to cut them on the ground or across two saw horses. To do this, set the posts loosely on the piers, brace and level them, and mark them as described above. Remove the posts, cut them, relevel them and fasten them in place.

To level posts on a freestanding deck, start with the corner post on the lowest ground. Measure up the post to the cutoff point and mark it. From this mark, run leveled strings across remaining post tops, mark, and cut.

Two Ways To Determine Post Height

If the joist tops will be flush with the ledger, use this method: Attach a level chalk line between one corner post and the ledger end as shown. Snap line across faces of intermediate posts. Measure down from these marks the combined thicknesses of joist and beam and mark this location for cutoff.

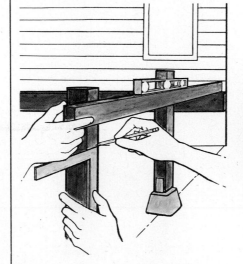

Use a try square or combination square to make cutoff marks around posts. Carefully cut off post tops with a portable circular saw.

If the joists will rest on top of the ledger, use this method: Extend a straight 2x4 from the top of the ledger to the corner post. Level the 2x4 with a carpenter's level. Mark where the bottom edge of the 2x4 intersects intermediate posts. Measure down from these marks the thickness of the beam, then mark and cut posts at this location.

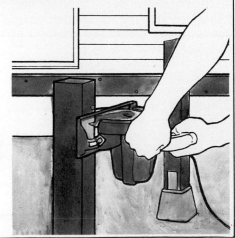

STEP 7—ATTACH BEAMS TO POSTS

There are different methods of attaching beams to posts. The easiest, toenailing at an angle through four sides of the post, is poor construction practice. The posts or beams may split and weaken the structure. Toenailing does not resist beam twisting.

Cleats made of 1x4s or 1/2- to 3/4-inch exterior plywood can be nailed to both sides of the posts and beams. The fasteners shown below make the strongest connections.

Design alternatives include using double posts, such as two 2x6s, with a single beam between them. Double beams bolted to a post may also be used. See drawings below. When the double-beam design is used, the post tops must be protected with metal flashing or wood cleats. Add these before laying deck boards. Where a single small beam is used with a larger post, the beam can be bolted to the post side. An alternative is to set the beam on the post and secure it with an angle iron. Flash exposed post tops.

Lay the beam across the post tops. Most lumber has a slight bow to it. Sight down the beam. Turn the beam so the bow side, or crown, is facing up. The slightly-bowed beam will then be slightly higher in the middle than at each end. Also place joists in this manner.

Check the beam at several locations along its length with a 2-foot carpenter's level. If the beam isn't level, use wooden shingles between the post and beam to shim up the low spots before attaching the beam to the posts.

Post-To-Beam Attachments

T-STRAP

DOUBLE BEAM BOLTED TO NOTCHED POST

METAL FLANGE AND NOTCHED BEAM

DOUBLE BEAM BOLTED TO POST

DOUBLE POST BOLTED TO BEAM

METAL CLEAT

METAL STRAPPING

ANGLE IRONS

SINGLE BEAM BOLTED TO POST

Splicing Beams—If you need to splice two beams together, make sure the splice is centered over a post top. Two methods of splicing beams are shown in the drawings below. If more than one splice is needed, stagger splices so they do not all fall on one line of posts.

Posts that support beams can also be used to support railings or benches. The beam is bolted to the side of the post. The post extends up through the decking to support the railing or bench. Protect the exposed post top with a *cap board* or *cap block*, as shown in the drawing on page 154.

STEP 8—ATTACH JOISTS TO BEAMS

Joists placed on top of beams may be toenailed or attached with metal fasteners. For joists attached between beams, use metal joist hangers as shown in the drawings on the facing page. You can also connect joists to beams by means of a cleat and angle irons, as shown in the drawing. Joist hangers can be used to connect joists to other joists or blocking to joists. If you toenail the joists to the beams, be careful to avoid splitting the joists.

To attach joists, start at one end of the deck. Mark the end joist location on the end beams, or the end beam and the ledger. Snap a chalk line across beam tops between the two marks. Use intermediate marks as a guide for attaching joists to intermediate beams. This will ensure straight joists. When attaching the joists, square them to end beams or ledger and end beam with a framing square.

Splicing Joists—It is not always possible or practical to buy joists in the lengths you need. Sometimes, you may have to splice two joists together. Joist splices should *always* be centered directly over a beam. Ideally, a joist splice should also be centered over a post top.

The drawings at right show two ways to splice joists. The first method involves using a wooden or metal cleat. For a stronger splice, you can use *gussets*, as shown in the drawing at left. Toenail an 8d nail through the tops of the joists at the splice. This will draw the ends together. Next, position and toenail the joist ends to the beam, then attach cleats or gussets on both sides of the splice. Cleats or gussets should overlap each side of the splice by 1 foot and be securely nailed or screwed.

The second method involves overlapping joists and nailing them together. Cut both pieces so they overlap the beam by at least 6 inches. For 2-inch-thick joists, drive 20d nails completely through both pieces and clinch nail ends. Alternatives to nailing include bolting or screwing pieces together. Joist splices should be staggered so they don't fall on the same beam.

Bracing Joists—Joists with spans over 8 feet should be cross braced to keep them from twisting. The wider the joist, the more likely it is to twist and the more bracing it will need. Cross bracing stiffens the structure

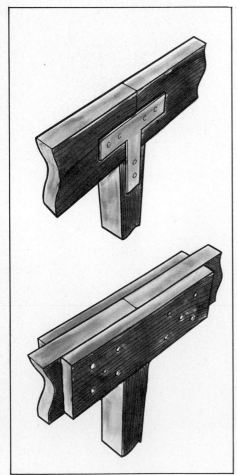

Beams can be spliced with a T-strap, top, or wood gussets, bottom. Beam splice should always be centered over a post.

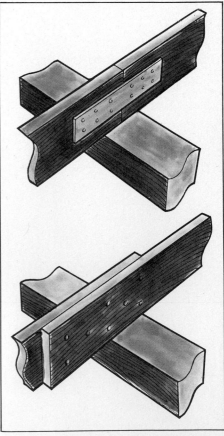

Joists can be spliced with a wooden or metal cleat, top, or overlapped on the beam, bottom. Joist splices should be centered directly over the beam. Overlapped joists should extend past the beam 6 or more inches.

Joist-To-Beam Attachments

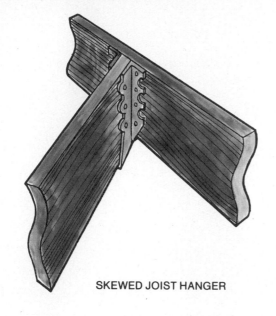

SKEWED JOIST HANGER

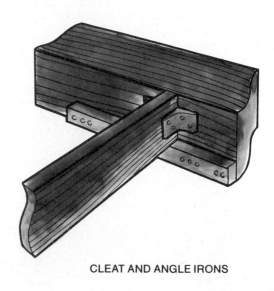

CLEAT AND ANGLE IRONS

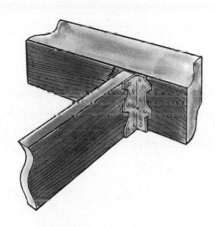

FRAMING ANCHOR

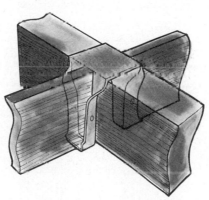

DOUBLE JOIST HANGER

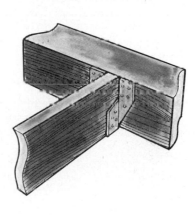

HURRICANE ANCHOR

SINGLE JOIST HANGER

TOENAILING JOIST
TO TOP OF BEAM

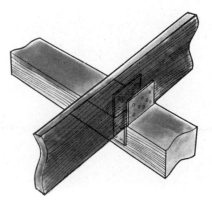

BEAM SADDLE ANCHOR

and distributes its *live load*. Live load is the weight placed on the deck surface. See page 125. The easiest way to brace joists is to install spacer blocks cut from the same size lumber as the joists. Stagger blocks so they can be end nailed through the joists. Space braces 8' apart along the joist span.

For joists 10 inches or wider, spacer blocks cut from the joist stock will add unnecessary weight to the deck. For wider joists, *cross bridging* is usually more practical and economical. Cross bridges can be of wood or metal, as shown in the drawings below.

For joist spans of 8 feet or less, *headers* nailed across joist ends provide sufficient bracing. Headers may be used to cover exposed joist ends, no matter what the span. See drawings below. This gives the deck a finished appearance. If the deck boards overhang the end joists, you can attach *fascia boards* to them to match the adjacent headers. Headers and fascia boards protect cut ends of joists and deck boards from decay. They also help hide unsightly substructure members on low-level decks.

Joist Detail

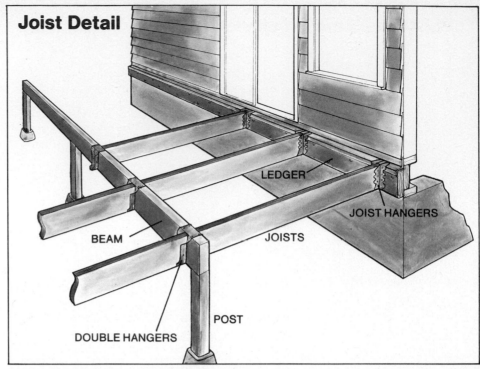

Joists are fastened to beam with double joist hangers, to ledger with joist hangers. In this drawing the joist tops are flush with the top of the ledger. The deck boards will run perpendicular to the joists, flush with the bottom of the door sill.

Joist Bracing

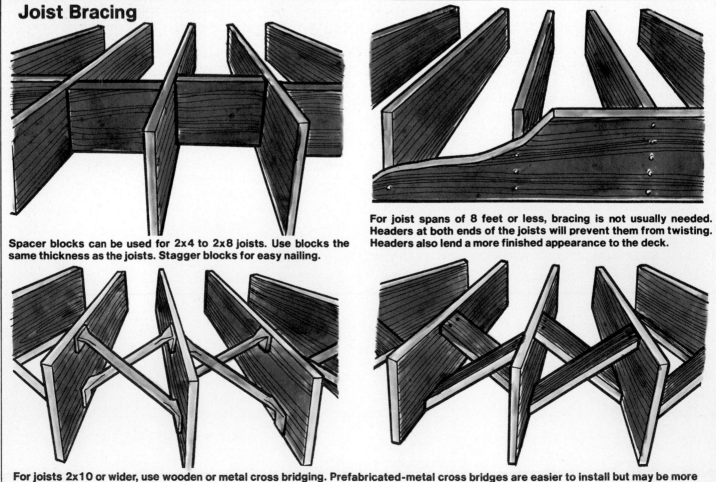

Spacer blocks can be used for 2x4 to 2x8 joists. Use blocks the same thickness as the joists. Stagger blocks for easy nailing.

For joist spans of 8 feet or less, bracing is not usually needed. Headers at both ends of the joists will prevent them from twisting. Headers also lend a more finished appearance to the deck.

For joists 2x10 or wider, use wooden or metal cross bridging. Prefabricated-metal cross bridges are easier to install but may be more expensive than wood bridging.

Substructure Bracing

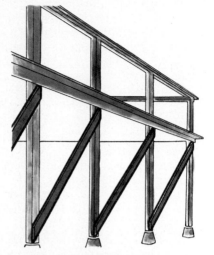

SINGLE DIRECTION BRACING

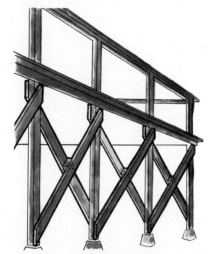

"X" BRACING WITH BLOCK SPACERS

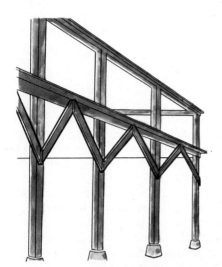

"Y" BRACING

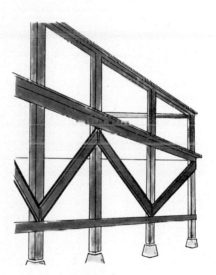

"W" BRACING

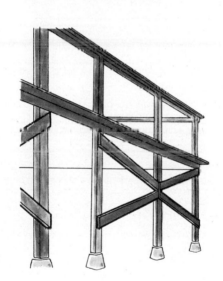

"K" BRACING

Leave 1/4" gap for drainage

"Y" DETAIL

"K" DETAIL

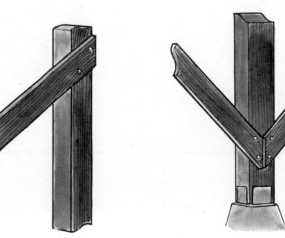

"X" DETAILS

"W" DETAIL

STEP 9—BRACING THE SUBSTRUCTURE

For decks 4 feet or higher, posts should be cross braced to resist lateral movement. Use 2x4s for braces up to 8 feet, 2x6s for braces 8 feet and longer. As a rule you only need to brace the posts around the perimeter or the deck. Bracing should be continuous around the perimeter of the deck. Select one of the bracing patterns shown in the drawing on page 149. Be sure to allow access under the deck. Trim brace ends flush with posts and beams.

Joining the braces must be done carefully to avoid trapping moisture. Washers between posts and bolted braces will allow space between pieces for air circulation.

STEP 10—FASTEN DECK BOARDS

Deck-board patterns may be parallel, diagonal, parquet or some other design of your choosing. Patterns can be enhanced by alternating decking widths, using any combination of 2x2s, 2x4s and 2x6s. Placing 2x4s on edge creates a tighter pattern.

Weeds growing under a deck can contribute to decay of posts, beams, and other substructure members. Before you nail down the deck boards, soak the ground underneath the deck with a weed killer. Then cover the ground with 4- to 6-mil. polyethylene sheeting or 30-pound asphalt roofing felt. Punch a few holes in the covering to allow water to drain through it. You may want to treat the ground with an insecticide at this time. This will kill any termites or other pests that may reside in the soil.

To lay deck boards, start at one end of the deck. Pick the straightest piece and square it to the end joists. Snap a chalk line across joist ends to make sure they are all square. Align the deck board with the chalk line. If some of the joist ends stick out past the end deck board, cut them flush to it.

If the deck has a ledger, leave a 1/4-inch gap between the deck and house wall.

If deck boards do not span the full width of the deck, cut boards so their ends are centered over intermediate joists. In other words, all board ends should be supported. Alternate the board lengths so all splices do not fall on the same joist.

Lay deck boards with the bark side up. See "Lumber Cuts" pages 129-130. Use 16d common nails as spacers between deck boards. This spacing allows drainage and provides a safe walking surface. Cracks can be easily cleaned with an ice pick.

Nailing—When nailing deck boards to joists, drive nails at an angle, as shown in the drawing below. Use rust-resistant nails for the deck surface. See page 135. If deck boards are 2x6s or wider, use three nails at the ends of each board. To avoid splitting board ends, flatten the nail points with a hammer before driving them. Place nails at least 1 inch in from board ends.

Laying 2x4s on edge rather than flat will give you a stronger deck. It will allow the pieces a longer span between joists. Toenail the pieces at each joist.

If boards are slightly bowed, place them bowed-side up. Otherwise, lay boards with the best-looking side up. To keep on-edge boards from twisting, use small 1/8-inch-thick strips of hardboard for permanent spacers. Cut spacers the same thickness as the deck boards—for 2x4s, 3-1/2 inches—and to equal length. Center spacers over the joists.

As you lay successive boards onto the joists, some won't be as straight as others. You can straighten some crooked boards as you nail them. To allow even spacing, fasten the board to one of the end joists with one nail. Go to the next joist and use a wood chisel to pry the board into position, as shown in the center drawings on the facing page. Fasten the board to the joist with one nail. Apply pressure with the chisel as you drive the nail. Continue this process on successive joists, using one nail to fasten the board and adjusting the spacing as you go. When you've finished, drive a second nail through the board at each joist location, angling the nail in the opposite direction to the first.

When you've finished nailing all of the deck boards, cut the board ends flush. Use a chalk line or straight board and pencil to mark board ends for cutting. Cut board ends with a

Nailing Deck Boards

Use 16d nails as spacers between deck boards.

To keep boards from working loose, drive nails at opposite angles.

Laying Deck Boards

Use a nail set to sink nails below the board surface.

Deck boards on edge are spaced with short lengths of wood or hardboard. To hide nail heads, drive nails as shown.

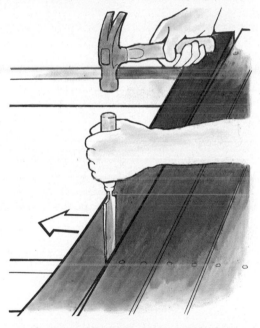

If a board bows inward, use a hammer and large chisel or pry bar to pry the board out to the correct spacing.

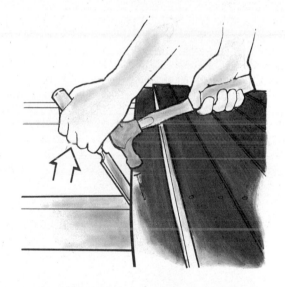

If a board bows outward, pry it inward at the joist to the correct spacing. Drive nail at angle shown to help force the board into position, then drive a second nail at the opposite angle.

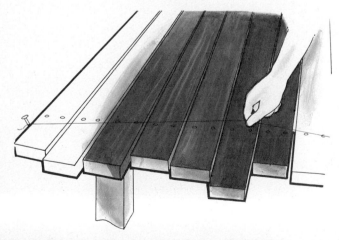

Use a chalk line to snap a straight line along deck edge.

Cut along chalk line with a portable circular saw. You can use a long board fastened to the deck with C-clamps to help guide the saw.

Deck Shapes And Board Patterns

Decks are built in many shapes and deck boards are laid in many patterns. Keep in mind that the deck substructures must be designed to accommodate the board pattern you've chosen. Here are eight deck variations that show deck-board patterns and the substructure needed to support them.

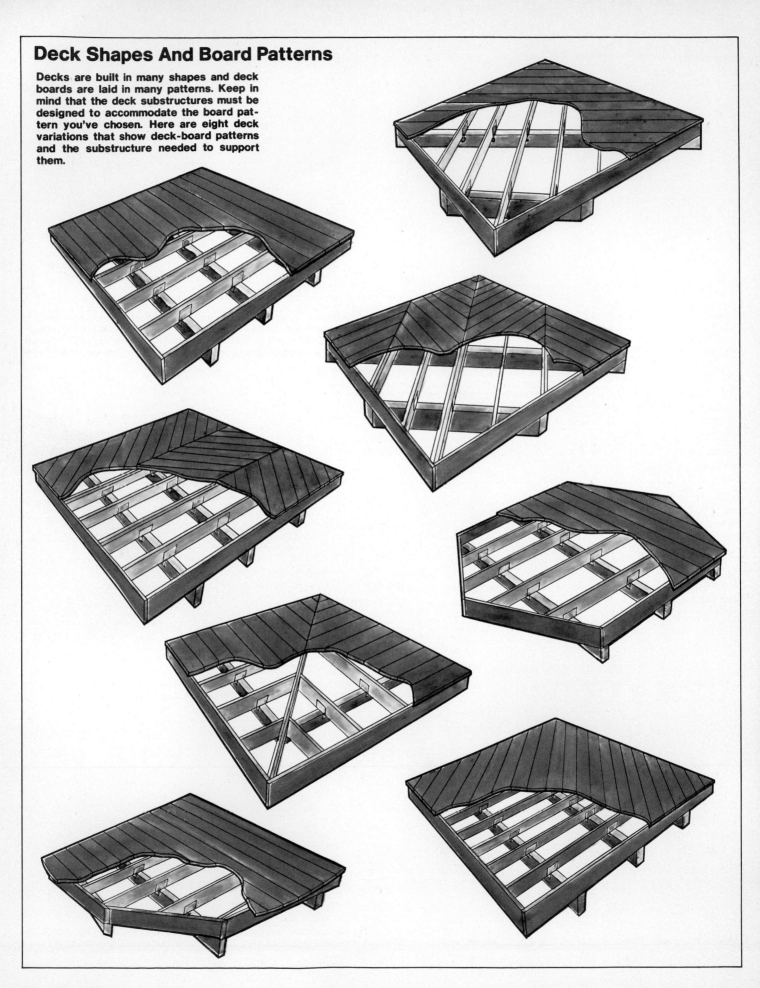

portable circular saw. See drawing on page 151. Treat cut ends of deck boards with a preservative, then cover with a fascia board as described on page 148.

Use a nail set to set nails slightly below the deck surface. This not only keeps nails from working loose, but helps keep nail heads from staining the surface. Protruding nail heads are dangerous on any walking surface. If you don't want nail heads to show, set the nails about 1/4 inch below the deck surface and fill the nail holes with exterior wood putty.

Custom Features For Your Deck

One way to customize a deck is to lay deck boards in interesting patterns, as shown on the facing page. Another way to give the deck a finished appearance is to attach headers and fascia boards around the deck's perimeter, as described on page 148. But the real opportunity to customize your deck comes with the addition of railings, benches, and steps.

RAILINGS

Railing designs seem limitless, but there are some controlling factors. Usually, if the deck is more than 3 feet off the ground, the railing must be at least 3 feet high. It should be sturdy enough to support people sitting or leaning on it. A railing should have no opening greater than 8 or 9 inches. This discourages children from crawling through. If you choose a solid railing for privacy, leave a 4-inch opening at the bottom. The opening eases sweeping and hosing the deck off. Check local codes for railing requirements.

When attaching railings, avoid creating joints where moisture can collect. This speeds decay and shortens the life of the deck. Secure railing posts to the deck with lag screws, bolts, or metal connectors. These fasteners hold much better than nails. Posts used for supporting the deck can also be used as railing or bench supports. For details, see pages 154-155.

There are several methods of attaching posts for railings as shown in the drawings on page 154. Posts can be bolted to a joist, fascia board, or header. Space posts about 4 feet apart

Fitting Deck Boards Around Rocks

To fit deck boards around rocks or other natural objects, you'll need a *scribe*, as shown in the drawing. Butt the board against the rock and use the scribe to transfer the rock contour onto the board. Use a portable jigsaw to cut along the scribed line. Angle the cut to conform to the vertical curve of the rock, as shown.

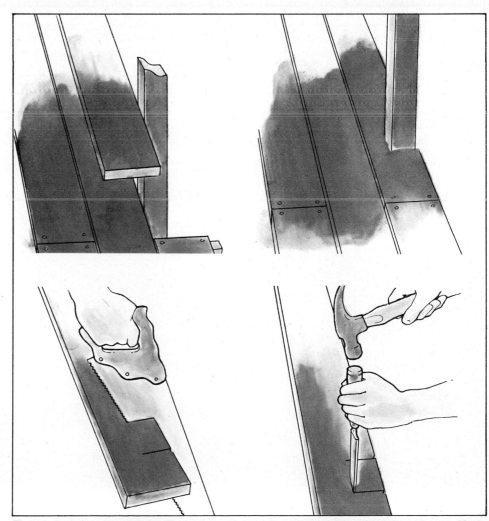

To cut deck board around railing post, set board against inside of post as shown. Mark width and depth of post on board, then add 1/4" space for clearance. Use a handsaw and chisel to cut the notch in the board.

Baluster Railings

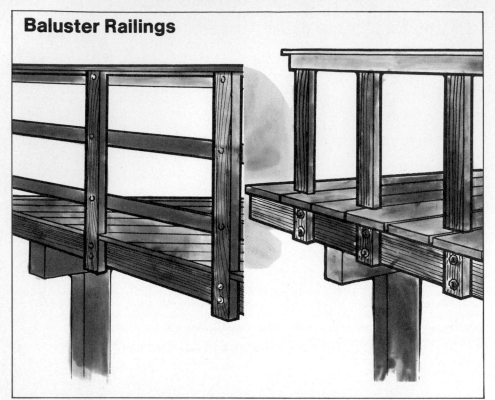

Baluster railings are generally used on decks 5 or more feet above the ground. On the railing at left, widely spaced vertical members require crossrails to help keep small children from climbing through the railing. On the railing at right, vertical members are spaced no more than 10" apart.

Cap Boards

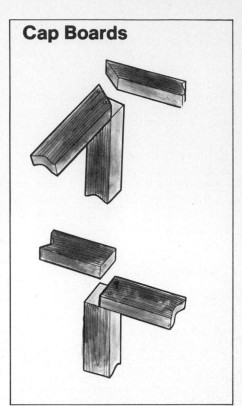

Cap boards are attached to post tops with nails or wood screws. Board ends are mitered at corner posts, top, or butted directly over intermediate post tops, bottom.

Posts used to support decks can be extended to serve as railing posts. Bolt the extended posts to the joists or beams, or to bracing between them.

Attaching Rails

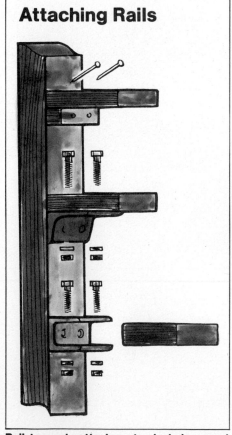

Rail-to-post attachments include wood cleat, top, angle iron, center, and metal connector, bottom.

to support horizontal railing pieces. Railing posts can also be spliced to deck posts. Splice them together between deck boards or joists, or support either side of the splice with blocking run between joists or beams. If beams or joists extend past the decking boards, pairs of railing posts can be connected to the joist or beam ends.

Instead of posts, you can use smaller vertical members, usually 2x2s, to support the railing. These members are spaced about 10 inches apart to form a *baluster railing*. See drawings above.

Several ways to fasten rails to posts are *wood block supports, metal connectors* and *angle irons*. Setting rails into posts with a notch or dado joint makes a neat-looking installation, but the joints have a tendency to trap moisture. Cap boards or cap blocks protect the exposed post tops from moisture. Cap boards may be nailed directly to the posts or attached with angle irons. The latter method is better. If the cap board is set at an angle, it will discourage people from sitting on the railing and provide drainage.

Before fastening railing members, treat cut pieces and any inset joints with a wood preservative. Angle the

Bench Details

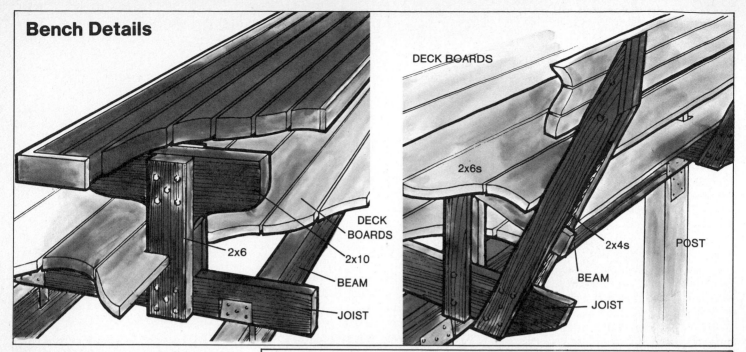

DECK BOARDS

2x6

DECK BOARDS

2x10

BEAM

JOIST

2x6s

2x4s

POST

BEAM

JOIST

Left: Bench supports are attached to deck joists. Vertical members are 2x6s. Horizontal 2x10 is cut with a portable jigsaw to add a decorative touch. Above Right: Joist ends protrude past deck edge to provide support for bench backrest.

cuts to allow drainage from joints. Leave about a 1/4-inch space when cutting deck pieces to fit around rails. Leave a 1/2-inch space between rail ends where they meet. This allows the ends to dry quickly after a rain.

Common railing designs and details are shown in the drawings on the facing page.

BENCHES

Benches around a deck are functional. They often serve as a railing as well as a place to sit. They can add a decorative element as well. On high decks, railings can serve as the backs of benches. The railing cap board can provide a place to set plates or glasses. On low decks, benches can be built without backs to avoid blocking views. They can provide a display area for potted plants or other items. On a large deck, wood tables and benches can be built in or left freestanding for flexibility of use.

Two bench details are shown in the drawings above.

STAIRS

Most decks will need at least a step or two. Others will have stairs leading from one level to the next. On sloping sites, you may want to add a series of smaller decks or landings to the main

Stair Detail

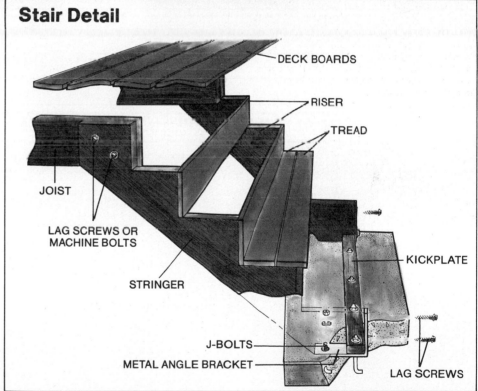

DECK BOARDS

RISER

TREAD

JOIST

LAG SCREWS OR MACHINE BOLTS

STRINGER

KICKPLATE

J-BOLTS

METAL ANGLE BRACKET

LAG SCREWS

deck, connected by steps.

Allow adequate drainage around step areas. Be careful not to create areas that will catch water runoff behind steps. Use decay-resistant wood for all step members.

Stairs are built on *stringers*, also called a *carriage*. These support the stair treads. See drawing above. Pairs of stringers should be spaced no more than 3 feet apart. Stringers are usually cut from 2x10s or 2x12s. They should be firmly fastened to the deck and to a

concrete *stoop* at the bottom of the steps. Depending on the stair design, you can either leave the vertical spaces between treads open, or close them in with *risers*. See drawings on page 156.

To determine number of steps needed, correlate tread width, or *run* with the step height, or *rise*. For easy-to-climb steps, the tread width in inches plus twice the riser height in inches should equal 23 to 25 inches. For example, if the tread is 12 inches wide, the step rise should be 6 inches.

Before you build stairs or steps, check local building codes. They often specify minimum tread widths and riser heights. Most codes require railings on stairs over a specified height, and that steps be consistent in rise and tread width within a flight, or run of stairs. For more information on designing stairs, see page 27.

There are several ways to attach treads to stringers. One method is to make sawtooth cuts in the stringers and set the treads into the angled cuts. Another is to cut channels or *dadoes* into the sides of the stringers and inset the treads between them. A variation of this is to attach cleats or angle braces to the stringers and attach the treads to them. Slope the cleats slightly downward to allow for drainage.

When placing treads, place the bark side of the lumber up. If you don't know how to tell the bark side of a board, see pages 129-130. A double tread of 2x6s or 2x8s is better than one wide piece. They allow better drainage and are not as likely to warp.

Stairways of more than one or two steps should have railings. Design stair railings to tie into deck railings both physically and designwise.

STANDARD TREAD-RISER RATIOS

Tread Width	Riser Height
11"[1]	6-1/2"
12"[2]	6"
13"	5-1/2"
14"	5"
15"	4-1/2"
16"	4"

[1]Treads less than 11" wide are not recommended.
[2]Most useful tread-riser ratio for decks.

Ramp Detail

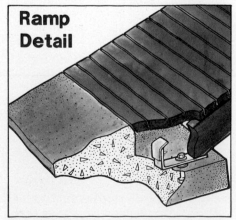

Concrete stoops for ramps must be specially designed to provide a smooth transition from wood ramp to ground level. Design may require special hardware. Seek help from an experienced designer.

Two Step Designs

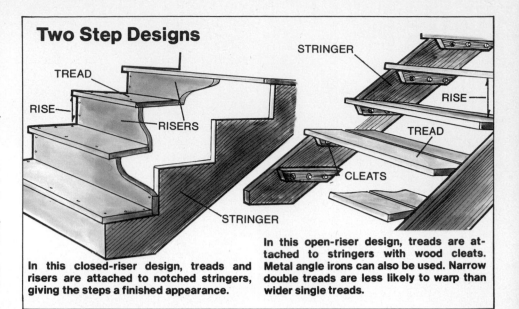

In this closed-riser design, treads and risers are attached to notched stringers, giving the steps a finished appearance.

In this open-riser design, treads are attached to stringers with wood cleats. Metal angle irons can also be used. Narrow double treads are less likely to warp than wider single treads.

Attaching Stringers

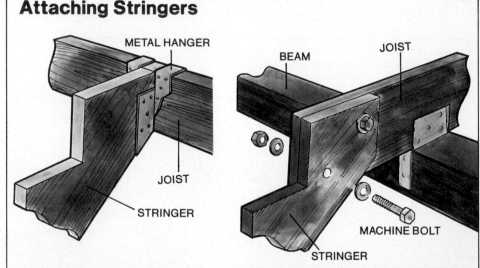

Left: If joists will run perpendicular to stringers, use a metal connector to attach them. Right: If joists will run parallel to stringers, extend joists past beams and bolt stringers to them.

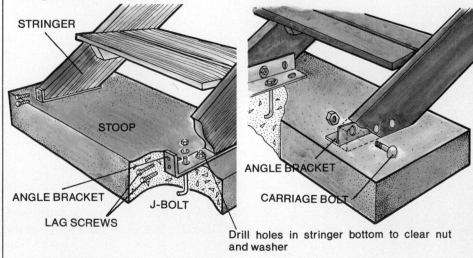

Drill holes in stringer bottom to clear nut and washer

Stringers must be securely attached to a concrete pad, called a *stoop*. These drawings show two ways of attaching open-riser steps. Flashing protects stringer ends from moisture buildup. See the drawing above for closed-riser detail.

Wood Deck Refurbishing And Maintenance

Careful construction, application of preservative and proper drainage where wood is joined will help ensure a long-lived deck. Regular maintenance is vital for an extended life.

Even with proper spacing of deck boards, leaves, dirt and other refuse will accumulate in the cracks and trap moisture. This leads to decay. A deck is also subject to the constant weathering effects of sun, rain and snow. The wood shrinks and swells as moisture changes. This sometimes twists and loosens nails. Added wear comes from simply using the deck.

Keep cracks between deck boards clean. A good maintenance schedule includes thoroughly cleaning cracks in early autumn just after the leaves have fallen, again in late autumn, and again in the spring just before you start using the deck for summer activities. Regular sweeping and hosing the deck surface will help considerably.

If deck boards are spaced properly—about 3/16 inch to 1/4-inch apart— clean cracks with a portion of straightened clothes hanger or an ice pick. If the deck boards are spaced tightly, use a putty knife to clean between them. A brisk hosing afterwards will wash out any debris left after cleaning.

While cleaning the cracks, check for boards that are loose, badly warped or are beginning to rot. Renail loose boards and replace defective ones.

When you replace rotted deck boards, check all adjoining wood to see that the rot has not spread. Apply wood preservative to exposed joists or beams and to new boards before joining pieces. Protruding nails can be countersunk with a nail set, as shown in the photo, top right. Setting nails during construction will prevent them from staining the deck surface.

Keep the deck surface clean and free of mildew. Shaded areas on the deck are especially susceptible to mold and mildew. Mildew-causing fungus may first appear as dark spots easily mistaken for dirt. They gradually fan out and become large areas of black or brown. You can control mildew by regularly scrubbing the deck with solutions described here. For serious mildew problems, treat

1. Use a putty knife to clean out cracks between deck boards.

3. Scrub deck with broom and TSP-bleach solution.

5. Use an oxalic acid solution to help restore natural wood color. Wear rubber gloves for protection.

2. Replace decayed or damaged deck boards if necessary. Countersink protruding nails with a nailset.

4. Use hand brush to reach difficult areas.

6. Use a long-handled paint roller to apply a water-repellent sealer or stain.

wood with a commercial mildewcide. Some wood finishes contain a mildewcide as well as a wood preservative. Consider this type of finish when you paint or stain the deck.

For scrubbing large areas or the whole deck, use a stiff broom or scrub brush. Make a solution of trisodium phosphate (TSP) and bleach—a cup of each to a gallon of water. Rinse with a solution of half water and half bleach. Use rubber gloves when applying solutions. Follow this procedure once a year—more often for stubborn, shady areas.

If your unstained deck has aged, you can restore some of its natural color with an oxalic acid solution. The solution should not be applied in full sun. The longer it takes to dry the more effective it will be.

First, clean the wood as described above. Put on your rubber gloves. Mix 4 ounces of oxalic acid with a gallon of hot water in a glass or plastic container. Slowly pour the acid into the water. Do *not* pour the water over the acid.

Apply the solution to wood with a cloth or brush. After the solution has dried, rinse the deck thoroughly with water. Stubborn pitch or rust stains can be removed with a second application of the acid solution.

If you like the appearance of weathered wood, or you want a new deck to match the color of surrounding weathered structures, you can speed the weathering process. Coat the wood with a solution of 1 part baking soda to 5 parts water. Let stand overnight and rinse with water.

Index

Sources Of Helpful Information

MASONRY PRODUCTS

Brick Institute of America
1750 Old Meadow Road
McLean, VA 22101
(703) 893-4010
Information on brick patios.

Portland Cement Association
5420 Old Orchard Road
Skokie, IL 60077
(312) 966-6200
Information on concrete patios and foundations.

LUMBER

California Redwood Association
One Lombard Street
San Francisco, CA 94111
(415) 392-7880
Redwood deck plans and ideas.

Council of Forest Industries of British Columbia
1055 West Hastings Street, Suite 1500
Vancouver, B.C. V6E 2H1 Canada
(604) 684-0211
Ideas for wood decks.

Southern Forest Products Association
P.O. Box 52468
New Orleans, LA 70152
(504) 443-4464
Information on use of Southern pine for decks.

Western Red Cedar Lumber Association
1500 Yeon Building
Portland, OR 97204
(503) 224-3930
Deck ideas and tree species information.

Western Wood Products Association
1500 Yeon Building
Portland, OR 97204
(503) 224-3930
List of deck plans, idea booklets and wood species information.

PLYWOOD

American Plywood Association
P.O. Box 11700
Tacoma, WA 98411
(206) 565-6600
Information on outdoor uses of plywood and plans for patio and deck projects.

WOOD PRESERVATIVES

American Wood Preservers Institute
1651 Old Meadow Road
McLean, VA 22101
(703) 893-4005
Information on pressure-treated lumber and do-it-yourself wood treatment.

Western Wood Preservers Institute
P.O. Box 748
Del Mar, CA 92014
(714) 455-7560
Information on pressure-treated wood.

HPBooks On Related Subjects

Fences, Gates & Walls
by S. Chamberlin and J. Pollack

Gardening in Small Spaces
by Jack Kramer

Hedges, Screens & Espaliers
by Susan Chamberlin

Lawns & Ground Covers
by Michael MacCaskey

Plants for Dry Climates
by Mary Rose Duffield and Warren Jones

Southern Home Landscaping
by Ken Smith

Spas & Hot Tubs
by A. Cort Sinnes

Western Home Landscaping
by Ken Smith

5.712770927351